The Future of Emergency Medical Services

THE FUTURE OF EMERGENCY MEDICAL SERVICES

ARTIFICIAL INTELLIGENCE, TECHNOLOGY, AND INNOVATION

2ND EDITION

DONNIE WOODYARD JR.

FIRE ENGINEERING™ • BOOKS •

Disclaimer
The recommendations, advice, descriptions, and methods in this book are presented solely for educational purposes. Photos are for instructional purposes only. Always wear the proper level of approved PPE when conducting training drills and operating at incidents. The author and publisher assume no liability whatsoever for any loss or damage that results from the use of any of the material in this book. Use of the material in this book is solely at the risk of the user.

Fire Engineering Books
110 S. Hartford Ave., Suite 200
Tulsa, Oklahoma 74120 USA

800.752.9764
+1.918.831.9421
info@fireengineeringbooks.com
www.FireEngineeringBooks.com

President: Eric Schlett
Vice President/Editorial Director—Books: Diane Rothschild
Acquisitions: David Rhodes, Diane Rothschild
Senior Development Editor: Daniel Edward Petrino
Production Manager: Tony Quinn
Customer Success Manager: Lane Nash
Book Designer: Alexandra Bonin, TIPS Publishing Services, Inc.
Cover Designer: Brandon Ash

ISBN 9781593705299

Library of Congress Cataloging-in-Publication Data Available on Request

Printed in the United States of America

1 2 3 4 5   30 29 28 27 26

To my inspiring sister: Your unwavering dedication as an emergency department nurse and now as a nursing supervisor, spanning over 20 years, is a testament to your professional excellence. You have skillfully balanced your demanding career while raising a beautiful family of five children. Your resilience and commitment have been a beacon of inspiration to me and countless others.

Also, to my fantastic team of trailblazers in Sri Lanka: Between 2006 and 2013, I had the privilege of leading an extraordinary group of visionary leaders. Despite facing insurmountable challenges and obstacles, you embraced a vision for a better Sri Lanka—a vision for a national EMS system. Together, we achieved remarkable milestones: teaching the first EMT classes; writing the first textbooks in Sinhala and Tamil; creating training videos; establishing national standards; forming the first Registry of EMTs for the country; and setting up the first medical emergency phone number, dispatch center, aeromedical transport, and international standard ambulance service. Today, these leaders continue to innovate, lead, and implement a system that stands as one of the best in the world. Your dedication and perseverance have laid a foundation that will benefit future generations.

I also extend my most profound appreciation to the Southwest Texas Regional Advisory Council for inviting me to speak on the future of EMS at their 2024 regional conference. Preparing for this presentation motivated me to write this book.

Contents

Foreword

The oldest and strongest emotion of mankind is fear, and the oldest and strongest kind of fear is fear of the unknown.

—H.P. Lovecraft, American writer

Knowledge breeds confidence. Confidence destroys fear.

—Robert Kiyosaki, American author and businessman

In the first quarter of the 21st century, artificial intelligence stands as one of the most transformative forces in technology—evoking both excitement and caution. But with understanding comes confidence, and with confidence comes the ability to lead.

The participants in every corner of society—from politicians to government leaders, authors to artists, captains of industry to tech moguls, clinicians to healthcare leaders, and students to educators—must, whether they choose to or not, confront a world upon which artificial intelligence (AI) has rapidly descended. So swift has been the proliferation of AI that the world has barely been able to grapple with the myriad of implications that come with it. As often happens, technology is outpacing the policy implications it creates. Health care is no exception, and emergency medical services (EMS) are probably even further behind the AI curve than other components of the American medical system.

I believe that the two quotes above frame the issue of AI and future-oriented technologies as well as any. One need only look at healthcare history to see a long line of beneficial technologies that bred fear and required the confidence of knowledge to overcome.

The turn of the 20th century saw the dawn of the science of defibrillation, a technology which has saved countless lives in the 100+ years since it was first demonstrated in 1899. The thought of applying electrical current to a person led many to eschew the use of defibrillation—primarily out of fear for its consequences. Despite its clear and demonstrated benefits, to this day, research reveals that utilization of public access to automated external defibrillators is, in part, inhibited by fear.

The advent of the polio vaccine, in which strains of the actual polio virus are administered to inoculate a healthy individual against the development of the disease, required a massive leap of faith (and data) to overcome fear. That fear inevitably came with administering into a body the very virus that was dreaded by the world and had caused so much misery (including the paralysis of our 32nd president, Franklin D. Roosevelt). To this day, fear (and misinformation) contribute to a substantial proliferation of vaccine rejection in the population, with the attendant public health challenges that denialism causes.

Leaders of nuclear nations must grapple with the existential issue of AI's ability to destroy civilization. That is pretty heady stuff. But leaders in all fields must address questions which, while perhaps less weighty to mankind, are nevertheless crucial to other aspects of science, the arts, industry, and health care.

This is where Donnie Woodyard (again) does a service to EMS with *The Future of Emergency Medical Services: Artificial Intelligence, Technology, and Innovation.* This important work tackles AI and future-facing technology in EMS from both a macro and a micro perspective. The author pulls back the curtain on the evolution of AI policy in the lofty air of its most political levels, and then progressively zooms in so that the reader can see how it both effects and can benefit their EMS systems, their people, and their patients. Donnie's work allows EMS leaders, technologists, educators and practitioners to replace their fears with knowledge—and to build their confidence with that knowledge to help shape a future enhanced by the limitless potential of AI.

As a lover of history, my favorite aspect of this book is the way that Donnie weaves the threads of our past with the filaments of our future. *The Future of Emergency Medical Services* draws a straight line from our profession's greatest leaders and thinkers—and the ways they challenged the status quo and embraced progress—to the future unfolding before us. This book gives EMS leaders a road map on how to harness new technologies for good, rather than being consumed with the fear of the unknown and allowing that fear to drive technological paralysis. It all begins with knowledge and understanding. *The Future of Emergency Medical Services* is a must-read for anyone who aspires to help blaze these new trails.

When the history of EMS in the 21st century is written, with all the breathless change it entails, Donnie's name will be on the list of those who contributed to shaping it. Like Safar, Nagel, Page, and Caroline before him, the name Woodyard will undoubtedly take its place in the pantheon of this generation's EMS thought leaders. I enjoyed this book, and EMS owes Donnie a debt of gratitude for writing it. I hope others find it as insightful and revelatory as I did.

To the future!

—Douglas M. Wolfberg

Preface

Trying to predict the future is a discouraging and hazardous occupation... The only thing we can be sure of about the future is that it will be absolutely fantastic. So if what I say now seems to you to be very reasonable, then I will have failed completely. Only if what I tell you appears absolutely unbelievable do we have any chance of visualizing the future as it will actually happen.

—Arthur C. Clarke, 1964 World's Fair, New York

The Future Is Now—Why Emergency Medical Services Must Lead, Not Follow

In 1964, Arthur C. Clarke stood before a camera at the World's Fair and attempted something we still struggle to do today—peer into the future with clarity, knowing we will be wrong. He understood that the future would always appear absurd to the present, and he was correct.

The world he envisioned—filled with satellites, mobile devices, digital libraries of knowledge, and real-time voice communication across the globe—sounded laughable then. Today, it is ordinary, mundane, and expected.

That is precisely where emergency medical services (EMS) stands right now.

We are surrounded by technologies that would have seemed impossible from the back of a 1990s ambulance: machine learning, language models, autonomous vehicles, real-time biosensors, and cloud-based systems that promise seamless interoperability. These tools are no longer theoretical and they are not future promises. They are here, and already in service.

So, the only question that matters now is this:

Will EMS shape that future—or be shaped by it?

Emergency Medical Services at a Crossroads

In this book, I am not predicting that flying ambulances or robotic medics will take over EMS in the next decade. What I am saying is more immediate—and far more urgent. Everyone in EMS faces a critical decision:

Will EMS leaders, clinicians, and educators drive the profession's evolution through technology, or will we risk being replaced or marginalized by systems that advance independently?

The brutal reality is that EMS as a profession is already behind our medical and professional colleagues.

Across the country, EMS services are closing, shifts are going unstaffed, and attrition has reached alarming levels. Rural communities are left without coverage for hours, sometimes entire days. Clinicians are being asked to do more with less: less pay, less support, less documentation time, and less confidence that anyone is advocating for their safety or sustainability. The tools are aging, and the systems are patchworked. The profession is buckling under the weight of a complexity it was never designed to carry.

Our education system is also showing signs of strain. EMS employers increasingly expect new graduates to enter the field fully prepared for independent practice almost immediately. This expectation is unrealistic. While the scope and sophistication of EMS education have expanded significantly over the past few decades, the structure of our programs has not kept pace. We remain tethered to outdated models that measure readiness by classroom hours or seat time—an approach long since abandoned by other health professions. Efforts to add meaningful instructional time or align EMS credentials with academic degrees have met fierce resistance. The result is a workforce pipeline that is underprepared, overstressed, and unsustainable.

Meanwhile, the broader medical industry is accelerating. In 2023, Merck's Keytruda—the nation's best-selling medication—generated $25 billion in sales. To put that in perspective, one drug grossed more revenue than the entire EMS system billed nationwide. While EMS struggles to survive, the rest of health care is scaling through capital, innovation, and data.

At the same time, technology outside of EMS continues to advance—quietly, persistently, and with growing impact. Hospitals are now using artificial intelligence to write physician and nursing notes, interpret X-rays, schedule patients, and assist in diagnosing diseases. Drones are delivering blood to remote trauma scenes. Autonomous vehicles are transporting passengers through major cities without human drivers. Dispatch algorithms are outperforming human call-takers under pressure. Predictive models are shaping emergency response strategies to reduce mortality and increase efficiency.

Yet many EMS systems still operate on outdated platforms—software developed decades ago, constrained by pre-digital policies. Clinicians are expected to perform miracles in 10-minute patient encounters using protocols written for a different era. The gap between what is possible and what is practiced continues to grow. That gap is where disruption lives.

EMS is not the only healthcare profession under pressure to transform, but few are as structurally vulnerable. The profession lacks a cohesive national identity and operates without stable or sufficient funding. Most reimbursement remains tied to transportation rather than the actual delivery of medical care, reinforcing outdated, and misaligned incentives. Billing models vary widely, and external agencies typically determine payment policies. EMS also lacks meaningful legislative influence over the laws and regulations that shape its future. As a result, it remains dependent on other systems while bearing rising public expectations. With one of the highest attrition rates in health care and limited career advancement opportunities for most clinicians, EMS now stands at a critical inflection point. Today, that invisibility has become a liability.

Technology does not wait for tradition, and markets do not pause for consensus. The private sector is already investing in solutions to the most inefficient pain points in health care and EMS, whether they are prepared for this or not.

So the question returns:

Will we lead this transformation—
or become just another cautionary tale?

To be clear, these criticisms are not a eulogy for EMS. They are a rallying cry.

The same spirit that brought defibrillation into the field and turned basic emergency medical technicians into advanced life support clinicians is alive today. EMS has always adapted under pressure. What we lack is not capacity, but alignment, vision, and investment. This moment demands leadership, not fear, because those who understand what is coming can still shape what happens next.

This Is Not a Tech Book— It Is a Leadership Manual

The word "technology" causes many EMS professionals to shut down immediately. It can sound like another gimmick, system update, or tool someone expects us to trust, without giving us the time to learn it or a voice in how it is implemented.

That response is understandable.

Too many technology rollouts have overpromised and underdelivered. Too many EMS systems have been burdened with software no one likes and hardware no one maintains. Skepticism is not only reasonable, but also earned.

But this book is not about gadgets, coding languages, or the latest app. This book is about leadership.

It is about asking the hard questions that too few EMS agencies are willing to confront:

- What does clinical excellence look like in a world where AI can process more data than humans?
- How do we staff systems sustainably, ethically, and financially viable when the workforce we once relied on is leaving?
- How do we protect the heart of EMS—the compassion, presence, and human connection—when automation reshapes everything around it?

This book is not about replacing the EMS clinician. It is about equipping the EMS leader.

The uncomfortable truth is that artificial intelligence is already here; drones are flying and autonomous vehicles are already transporting passengers through major American cities. The only thing missing from many of these systems is us.

This book provides what EMS leaders need to navigate the future with purpose:

- A realistic picture of how AI, drones, and automation are already used in health care and emergency response.
- Case studies from real EMS systems—rural and urban, public and private—that are actively testing and adapting these tools.
- Ethical frameworks to guide decision-making when technology outpaces the protocol.
- Most importantly, you need a voice that respects you as a professional, not a user.

You do not need to speak Python or understand neural networks to care about the future of your profession. You need only the courage to lead through change. That is what this book is about.

What Is Coming Is Not Optional: Trajectory of Artificial Intelligence Versus Reality of Emergency Medical Services

Integrating AI into healthcare is no longer a matter of debate for EMS. This change is driven by federal policy, major hospital networks, venture capital, and evolving consumer expectations.

While many EMS agencies are still debating electronic patient care reporting vendors and relying on outdated scheduling software, healthcare's most powerful players already embed AI into diagnosis, documentation, resource allocation, and patient navigation. This is not fringe innovation. It is mainstream evolution.

Here is a reality check:

- Epic and Microsoft are integrating generative AI into hospital electronic health records nationwide.
- Amazon Clinic, Walgreens, and CVS use AI to triage patients, route prescriptions, and streamline clinical workflows—no ambulance required.
- The U.S. Department of Health and Human Services, the National Academies of Medicine, and the World Health Organization have each issued formal guidance supporting the large-scale adoption of AI in health care.

Meanwhile, EMS agencies across the nation are still doing the following:

- Documenting with click-box software designed in the early 2000s
- Following protocols that are not regularly updated and evidence-based
- Writing narratives that fail to protect the clinician when their actions are questioned

- Producing reports that do not contribute meaningful information to the broader healthcare system about the patient's condition at the time of injury or illness
- Relying on documentation practices that contribute directly to burnout, particularly in high-volume systems
- Requiring clinicians to complete narratives after their shift, sometimes hours later, because field conditions do not support focused, real-time documentation

AI does not share those limitations. It documents in real-time, adjusts to context, and generates clean, accurate narratives while the clinician focuses on care, not clicking boxes.

The gap between what is possible and what EMS is doing continues to widen. That gap is where disruption lives.

Think of what happened to Blockbuster, taxi services, and Kodak. These industries did not fail because they lacked infrastructure or expertise. They failed because they could not imagine a world that moved faster than their systems.

To be clear, I am not predicting that EMS will go extinct. People will always experience emergencies. They will continue to need care and transport. The need will not disappear, but the model will change—in fact, it *must* change.

What I am increasingly concerned about is not collapse but disruption. A transformation led from outside the profession. One that fundamentally reshapes how prehospital care is delivered, documented, and valued, without EMS at the center of that change. From a consumer standpoint, that kind of innovation might even be welcomed. However, as someone who has spent a lifetime in this profession—as a clinician, educator, regulator, and advocate—I believe it is far better for EMS to evolve from within than reengineered from without.

The future is not a distant prospect; it is an unfolding reality. EMS faces a stark choice: to shape this evolution proactively or to be passively reshaped by it.

This Book's Promise to You

This book is written for you—the EMS clinician, educator, supervisor, medical director, or system leader—who understands that we cannot continue doing things the way we always have and still expect to survive.

It is written for the people on the ground who know that burnout, understaffing, inefficiency, and outdated tools are not abstract problems. They are daily realities, and it is written with one clear purpose: to help you lead EMS through its most pivotal transformation since the 1970s.

In the chapters ahead, you will find the following material:

- Real-world examples of how AI, drones, automation, and data science are used now in EMS systems, hospitals, and global health networks
- Leadership frameworks to help you ask the right questions, anticipate downstream effects, and set policies that protect patients and providers
- Strategic guidance on piloting innovation, building trust with your team and community, and avoiding the costly mistakes made by other industries caught off guard by rapid change
- Ethical clarity on what we must protect—our humanity, our connection to patients, our accountability—and how to preserve those values in a world increasingly shaped by algorithms

You will not find hype here. Instead, you will discover pressure-tested insights, hear the stories of early adopters and quiet visionaries, and see our profession reflected in these pages, not the version that remains standing still.

This is not a technology manual. It is a call to action, because what comes next in EMS will not be led by software developers or national vendors. It will be led by you.

What Will EMS Be When the Dust Settles?

In every transformation era, a profession must decide what it will fight for—and what it is willing to let go of. For EMS, this is that era.

We are standing at the threshold of systems that process information faster than we can, vehicles that operate without human drivers, and algorithms capable of generating clinical documentation before we have even left the scene. That may sound unsettling. It should. However, it should also sound familiar. Because we have been here before.

> We were here when emergency transport shifted from funeral home hearses to mobile intensive care units.
>
> We were here when defibrillation moved out of the hospital and into the field.
>
> We were here when paramedics were first licensed, and standing orders replaced radio consults.
>
> We were here when 12-lead electrocardiograms, intraosseous infusion, and whole blood administration transitioned from hospital-based procedures to standard paramedic care.

Each of those moments was disruptive, each was met with resistance, and each ultimately expanded what EMS could do for the people we serve. Now we are here again.

The next decade will define what EMS becomes within the AI-enabled healthcare system. Will we shape the tools or learn to work around them? Will we lead with integrity—or scramble to keep up? Will we protect our profession's identity by evolving it, or cling to what is familiar until it fractures?

Because the truth is, this moment is not only about technology.

It is about control.

It is about relevance.

It is about survival.

More than that, it is about the kind of care we want to deliver in the future we are building now.

So, turn the page. Ask the hard questions and prepare to lead—not follow.

Acknowledgments

This book is the product of many conversations, shared risks, and collective belief in what emergency medical services (EMS) can become. While the ideas presented here focus on the future, artificial intelligence, and technology they are rooted in people, relationships, and lived experience.

I am deeply grateful to my friends and colleagues in Sri Lanka and across South Asia who believed in a vision of modern, professional, and patient-centered EMS. Through their courage, discipline, and sustained commitment they demonstrated that meaningful systems change is possible, even in the most complex environments. Their work did not simply improve services; it helped transform nations and reshape how EMS can be designed, governed, and trusted. Much of what I understand about building EMS systems at scale was forged alongside them.

I owe a profound debt to my colleagues and partners in Louisiana and Colorado, where I had the privilege of serving as State EMS Director. The clinicians, regulators, elected officials, healthcare partners, and community stakeholders in those states coached me, challenged me, supported me, and taught me what leadership under pressure looks like. Their willingness to collaborate, innovate, and remain patient-focused shaped my understanding of governance, accountability, and the practical realities that must guide any discussion of technology or artificial intelligence in EMS.

I am especially thankful to my colleagues at the National Registry of Emergency Medical Technicians, where I had the tremendous opportunity to serve both as chief information officer and chief operating officer. Together, we worked to redefine the role of data, technology, and user experience in professional credentialing. That work fundamentally influenced my thinking on how intelligent systems, ethical data use, and human-centered design can strengthen trust, competence, and mobility across the EMS profession.

To my family and colleagues in Giles County, Virginia, I offer my deepest thanks. It is often said that it takes a village, and in my case that was unquestionably true. Growing up in a community rich with mentors, role models, and quiet leaders instilled values of service, accountability, and humility that continue to guide my work. The lessons learned there remain foundational, even as the field evolves toward advanced technology and artificial intelligence.

Finally, thanks to my family for their patience, encouragement, and unwavering support. Their belief in this work and in the broader mission of advancing EMS made this book possible.

This book looks forward to what EMS can become through the thoughtful integration of technology and artificial intelligence, but it stands on the shoulders of people who believed early, worked hard, and chose to lead.

1

The Future Is Now

Why Emergency Medical Services Must Lead, Not Follow

LEARNING OBJECTIVES

After reading this chapter, the reader will be able to

- Explain the historical context of EMS development, focusing on the impact of federal disinvestment in the 1980s, and how this shaped EMS's fragmentation, workforce challenges, and skepticism toward national coordination.
- Understand the relevance of AI to EMS, including a practical definition of AI and its potential to address long-standing challenges like burnout, inefficiency, and documentation burdens.
- Distinguish between meaningful innovation and superficial technological upgrades in EMS, evaluating innovations by their ability to enhance clinical decision-making, improve patient outcomes, and streamline EMS operations.
- Identify emerging technologies already transforming health care and assess their potential applications within EMS, including AI-powered documentation, predictive analytics, real-time clinical decision support, and autonomous systems.
- Analyze the risks of inaction in EMS innovation, including the threat of becoming obsolete in a rapidly evolving healthcare landscape, and the consequences of not leading the integration of AI and emerging technologies.
- Recognize EMS's unique opportunity to shape the future of prehospital care, using emerging technologies to enhance system-wide performance, improve clinician well-being, and demonstrate the profession's value to the broader healthcare system.

My Story and the Things I Did Not Know

I grew up in a small Appalachian town nestled along the New River, tucked between the Virginia–West Virginia border and the winding path of the Appalachian Trail. It was quiet, remote, and deeply rooted in tradition. Our hospital had a four-bed emergency department, usually staffed by a family physician. The nearest trauma center was more than an hour away. I was a teenager—still in high school—when I first discovered EMS. If the TV antenna was angled just right, I could catch grainy episodes of *Emergency!* and watch Johnny and Roy rescue strangers with calm precision and a sense of purpose I could not look away from. Around the same time, *Rescue 911* hit the airwaves, dramatizing real emergencies and showing what it meant to show up when others could not. Those shows made a mark on me. I signed up for a CPR classes at the local rescue squad, and before long, I was spending more time in that cinder block station than anywhere else.

That rescue squad was filled with radios, oxygen tanks, worn leather jump kits, and a quiet pride that hung in the air alongside the scent of diesel and gauze. But it was not the gear or the sirens that hooked me—it was the opportunity to help people in need, and it was the people who made up the rescue squad. The EMS service was not just family-friendly, it was family-built and family-operated: one family in particular, the Davis family, *was* the agency. They trained the new volunteers, including me. They staffed the calls. They kept the place running. Their legacy *was* the organization. EMS, to me, looked like something sacred passed down through generations. But I did not know the history of the profession I had just entered.

This was the early 1990s. As a national system and profession, I had no idea that EMS was barely 20 years old. I did not know that in the 1960s and 1970s, Congress had passed not one, but two significant pieces of legislation aimed at building a national EMS system.

One, the Highway Safety Act of 1966[1], empowered the newly created U.S. Department of Transportation to lead EMS improvements, focusing primarily on traffic injury and rural trauma care. The Emergency Medical Services Systems Act of 1973[2] directed the Department of Health, Education, and Welfare to build comprehensive and coordinated EMS systems nationwide, with 15 essential components and substantial grant funding. Each federal agency had its vision, and both competed for influence, funding, and alignment from the states. That turf battle shaped the early architecture of EMS, but I knew none of it.

I certainly did not know what came next. In the early 1980s, amid a shifting political climate, rising economic pressures, and a nationwide push for deregulation, Congress withdrew targeted federal support for EMS. While no single law was formally repealed in 1980, the legislative groundwork was laid. By 1981, the Omnibus Budget Reconciliation Act had consolidated many categorical health programs, including EMS, into block grants to the states. The dedicated federal EMS agency was dissolved, and the national leadership effort abruptly ended.

The result? The dream of a unified, federally coordinated EMS system collapsed. What emerged in its place was not federal—it was hyperlocal. It was personal. It was heroic.

In towns like mine, people did not wait for a system—they became the system. They trained however and wherever they could. They borrowed gear, ran calls, and built the foundation of modern EMS from the ground up. They gave up nights and weekends, job security, and sleep—and they gave everything they had to get the job done.

As EMS matured, state governments were obliged to step in again, offering regulation and structure. However, something else had taken root by then: a deep skepticism toward oversight. A cultural memory of being abandoned by bureaucracy and left to survive alone. That memory shaped a generation of EMS leaders and still echoes today.

Many clinicians—especially those working in rural or volunteer systems—instinctively resist national standards or federal guidance. Not always because they can explain why, but because their trusted mentors modeled that resistance. It is not just about logic. It is about legacy and tradition, both of which run deep in EMS.

I do not share this to criticize our past. I share it to understand it. Because if we do not know where we came from, we will never understand why change is so difficult. Why innovation is resisted. Why is fragmentation tolerated? Why do so many still view EMS not as a national care system but as a local craft passed from generation to generation? We are not the untrained volunteers of the 1970s anymore. The public does not see us that way, and should not see us that way. We honor our traditions not by clinging to them but by ensuring they do not become the rock that trips us as we enter the future.

Nevertheless, understanding where we came from is no longer enough. Because today, EMS stands at another crossroads, and this time, the threat is not a lack of funding or federal direction—it is irrelevance in a healthcare system that is evolving without us. We are struggling with sustainability. We are struggling with the workforce. We are struggling with identity. Sometimes, it feels as though the only thing we all still agree on is this: we show up, care for the patients where they are, and get them to where they need to go. That mission is still clear. But everything around it is moving faster than we are.

Health care is not slowing down to wait for EMS. Artificial intelligence (AI) is already deployed in hospitals, clinics, pharmacies, dispatch centers, and now directly in the hands of patients through consumer health apps. Prehospital blood transfusion, once considered a military-only capability, is now being actively supported through federal National Highway Traffic Safety Administration and Department of Defense initiatives and is part of civilian EMS systems across the nation. Medications, diagnostics, and interventions that once took years to adopt are now fielded in months. In cardiac arrest, a highly trained paramedic team can now deliver care on par with any emergency department in the country.

We have kept pace with medicine, but EMS is falling behind in technology (fig. 1–1). If we do not catch up and attempt to lead the conversation, we risk

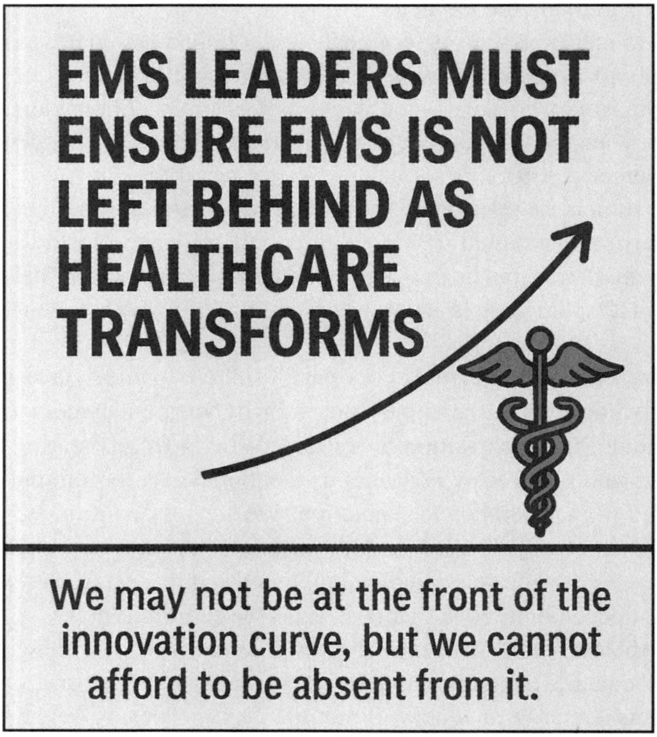

FIGURE 1–1. As health care rapidly evolves, EMS leaders must ensure the profession remains an active part of that transformation. EMS may not always lead the innovation curve, but absence from it is not an option.

becoming invisible again, just as it was before those first ambulances rolled out of cinder block garages in towns like mine.

This is not just about survival, nostalgia, or even leadership. It is about ensuring EMS is not left behind as health care transforms—because while we may not be at the front of the innovation curve, we cannot afford to be absent from it.

Now it's time for us to get to work.

Why Should I Care About Artificial Intelligence, Technology, and Innovation?

Being honest about where many EMS professionals stand on this topic is essential. If you are a clinician, educator, or supervisor, you likely did not enter this profession to chase the latest gadget. You did not sign up to learn Python. You did not join EMS to become a coder, data scientist, or digital futurist. You joined to help people, to care for them, to stabilize and save lives when minutes matter. And you are good at it.

That is why it is easy to tune out when conversations about AI, drones, automation, or machine learning begin. EMS has always found a way to deliver care with whatever resources and tools are available. We have built our success on adaptability, resilience, and ingenuity. If that has worked until now, why change?

Because the world around us is already changing, those changes are about redefining what EMS is, how we deliver care, and what it means to remain relevant within modern health care. Across the country, EMS is under scrutiny. States are launching sustainability studies. Multiple state are being reorganized. Legislators ask hard questions about cost, duplication, value, and outcomes. While EMS faces underfunding and workforce shortages, AI offers innovative solutions to alleviate these pressures and create a more sustainable future.

A 2024 media analysis by the American Ambulance Association and the Academy of International Mobile Healthcare Integration[3] reviewed more than 2,000 EMS-related news stories from January 2021 to June 2024. Nearly 80% of the coverage focused on workforce shortages and funding instability. More than 100 reports documented EMS system closures or agencies withdrawing from communities, often in rural or frontier regions. These statistics confirm what many already feel: The system is at a breaking point (fig. 1–2).

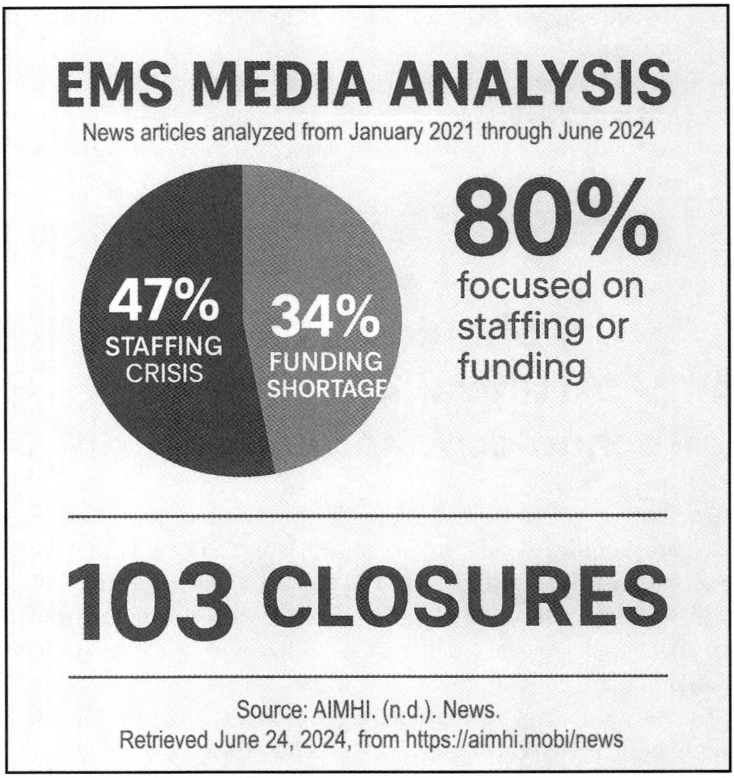

FIGURE 1–2. A nationwide review of EMS media coverage reveals a profession under strain: 80% of news stories published between 2021 and 2024 focused on staffing or funding crises, with 103 agency closures reported.

Several states have formally acknowledged the urgency of these challenges. In 2022, Colorado[4] established a five-year EMS System Sustainability Task Force to evaluate the long-term viability of its EMS infrastructure. Idaho and Minnesota launched similar initiatives, citing severe reimbursement gaps, workforce shortages, and unsustainable volunteer models. In South Dakota[5], the Department of Health identified seven major threats to EMS sustainability: workforce instability, system reliability, and insufficient financial resources. In Connecticut[6], State Senator Dr. Saud Anwar, vice chair of the Public Health Committee, gave the following warning:

> *There is a matter of time [until] we will have parts of our state and parts of our communities which will not have the EMS response that we would expect for our own family members...*

A local EMS chief[7] in Idaho echoed this urgency during 2024 legislative testimony:

> *With the current model for EMS in Idaho, it's not sustainable. If we do not do something to fix the system soon, it will completely fail in the next 5 years, and at most 10 years. What I mean by that is you're going to call 9-1-1 for yourself or your loved one and you're not going to get a response. Or you're going to get a very delayed response and it's a response from someone who is not medically trained.*

A 2023 official report from the South Dakota Department of Health[8] identified seven significant issues and challenges facing South Dakota's EMS system sustainability, including workforce shortages, system reliability, sustainability, and financial resources. A 2024 Minnesota EMS Task Force report[9] drew similar conclusions, stating the following:

> *EMS in Minnesota continue to face critical challenges to delivering adequate services across the state. Severe shortages in reimbursements, aging workforces, unsustainable volunteer staffing models, and increased operating costs are forcing stakeholders to come together to address these challenges confronting the future of EMS delivery and sustainability.*

These examples highlight that while not all EMS systems in the United States are struggling, many, particularly in rural and frontier areas, are grappling with significant sustainability issues, financial constraints, recruitment and retention challenges, and outdated operational models. Without innovative solutions, these communities risk the local EMS system collapsing, leading to delayed or nonexistent responses to emergency calls. AI and advanced technology have the potential to address these challenges and seize these opportunities.

At the same time, technology is advancing—without us. Hospitals are implementing real-time AI tools for triage and diagnosis. Retail clinics are using automated systems to deliver care without ever calling 9-1-1. Autonomous vehicles are transporting patients safely and reliably across major cities. Private companies are actively positioning to take over nonemergency medical transport using platforms that do not include EMS.

This is why you should care. The future of EMS will not be secured by tradition, necessity, or goodwill. It will be determined by how well we adapt, how quickly we lead, and how well we communicate what makes EMS indispensable—not just operationally but clinically, strategically, and technologically.

This chapter (and this book) will not teach you how to build the technology. It will teach you how to lead with it, evaluate it, advocate for tools that work, reject those that do not, and ensure that EMS remains at the table, not just as an end user but as a coarchitect of the future.

If you care about EMS as a profession—its sustainability, identity, and impact—then you already care about innovation, whether you realize it or not.

The American Nurses Association[10] emphasizes that "the appropriate use of AI in nursing practice supports and enhances the profession's core values and ethical obligations." Physicians have echoed this sentiment in clinical settings. "AI has allowed me, as a physician, to be 100% present for my patients," shared Dr. Michelle Thompson[11], reflecting on the impact of AI-powered documentation in her practice. These voices underscore a shared recognition: AI, when implemented thoughtfully, does not replace clinical care—it restores it by reducing friction, fatigue, and distraction.

What Is Artificial Intelligence (and What Is Not)?

To lead EMS into the future, we must begin with clarity. AI is often misunderstood, oversold by vendors, and underestimated by clinicians. However, to make informed decisions, we need a grounded understanding of what AI is, what it is not, and how it can be applied in real-world EMS settings.

AI is not magic, science fiction, or a single device, product, or platform. It is a broad category of computer systems designed to perform tasks that typically require human intelligence. These tasks include recognizing patterns, understanding language, making predictions, learning from data, and drawing conclusions based on probability rather than certainty.

Since 2023, access to AI tools has expanded dramatically. Like much of the public, many EMS professionals have encountered large language model platforms such as ChatGPT, Perplexity, etc. These tools have made AI more visible and, in some cases, tangible for the first time. However, while large language models have captured the public's imagination, they represent only one part of a broader, more impactful AI ecosystem.

In EMS terms, AI can be understood as any system that takes messy, real-world input—a patient complaint, a set of vital signs, or a field narrative—

and produces a judgment, recommendation, or action that meaningfully improves care. That is AI in its most practical form.

Importantly, AI is not a singular technology. It is a constellation of tools, each designed to address different types of problems. In EMS, these include systems for automated documentation, clinical decision support, dispatch triage, scheduling optimization, and more. Chapter 2 will explore the most relevant types of AI used in EMS today: what they do, how they work, and where the risks lie.

It is also essential to understand what AI is not. Unless your system refuses to adapt, AI will not replace your clinical judgment and will not take your job. It does not "know" anything in a human sense. It is not sentient, it does not feel, and it does not care. It operates through algorithms and statistical models—processing inputs, identifying patterns, and producing outputs at speeds humans cannot match. But it cannot interpret nuance, demonstrate compassion, or apply contextual reasoning the way a trained EMS clinician can.

Just as modern cardiac monitors can detect ST-elevation myocardial infarctions faster than most physicians, AI can deliver clinical insights faster than manual data review. It is not a replacement for training but a tool to extend and support it.

Understanding what AI is—and what it is not—is now a professional imperative. This foundational knowledge is the first step toward choosing vendors wisely, advocating for the right tools, protecting one's scope of practice, and leading innovation instead of reacting to it.

EMS clinicians who understand how AI works will be trusted to shape how it is used within their agencies, in state-level policy, and across the national profession. Those who do not will find themselves working in systems they did not help design, with no voice in how those systems operate, or leaving the profession altogether.

That can change. That must change. Now is the time to learn, lead, and act.

Emerging Technology: What Matters and Why Emergency Medical Services Must Pay Attention

AI is just one part of a larger wave of innovation. Emerging technologies, already transforming hospitals and public health, are also starting to reshape EMS. The question is not whether they will arrive. It is whether we are ready to use them wisely.

This readiness has already become reality in Sri Lanka, a nation of 23 million people. Over the past two decades, through a national collaboration supported by international partners, Sri Lanka has built what the World Bank described as "one of the world's most digitally advanced" ambulance programs, and the system continues to expand its use of AI and augmented reality in field operations. Clinicians routinely use augmented reality headsets for procedural guidance, view patient vital signs in a heads-up display, and conduct real-time consultations with remote physicians en route. AI algorithms analyze incoming data from monitors and assessments, offering real-time decision support and predicting potential deterioration. What was once futuristic is now field-tested. It provides a global example of how EMS systems, even in resource-limited settings, can lead with technology rather than wait to inherit it.

Emerging technology often sounds like a buzzword, but it carries a specific and practical meaning in EMS. These are not theoretical tools or futuristic concepts. These technologies are already transforming health care in adjacent fields such as emergency medicine, public health, and hospital operations. The only difference? They have not yet become standard in ours.

Emerging does not mean untested. It means in motion. It means these tools are being adopted, evaluated, and scaled. Far too often, EMS is the last to benefit. Sometimes, that delay is financial. Sometimes, it is regulatory. But often, it is cultural—a product of skepticism, outdated workflows, or fear of change. Regardless of the reason, the consequence is the same: EMS risks being left behind as the clinical environment evolves.

It is also important to be clear about what *does not* count as meaningful innovation. Not every new gadget or software update qualifies. Faster Wi-Fi, newer tablets, or tools that look modern but change nothing are not emerging technologies. In EMS, a tool must meet at least one of three criteria:

- Transform how EMS delivers care
- Improve how EMS integrates with the broader healthcare system
- Enhance how patients experience EMS—before, during, or after the call

If a technology does not meet these standards, it may be new but useless. True innovation reduces friction, solves real problems, supports the clinician, strengthens the system, and improves patient care. Anything less is noise.

Technology adoption is accelerating. The window between prototype and standard of care is shrinking. What once took a decade to implement now takes two years—or less. Systems that understand this are adapting. Those that do not are being restructured, replaced, or bypassed altogether.

EMS cannot afford to mistake novelty for progress. Knowing what to ignore is just as important as knowing what to embrace. AI is only part of a broader wave of emerging healthcare technologies. From real-time decision support to autonomous logistics, these tools are no longer futuristic—they are active, advancing, and rapidly spreading. Chapter 2 will walk through the ten categories of emerging technology most relevant to EMS and explain how they are applied today.

Want to know exactly how AI changes documentation, dispatch, and decision-making? Chapter 2 has ten real-world examples showing how AI is now reshaping EMS.

A New Tool, a New Opportunity

For decades, EMS has faced the same persistent challenges: workforce shortages, financial instability, fragmented systems, and professional invisibility. These problems are not new. What is new is the potential—an unprecedented opportunity to shift the trajectory of our profession, not by working harder, but by working smarter. For the first time, we have access to tools that can fundamentally alter how EMS operates. If used purposefully, these tools can help us do more than survive. They can help us redefine what EMS can become.

Since the early 1980s, when federal leadership was dissolved and national EMS development was left to the states without a roadmap or sustained funding, our profession has held the system together through grit, creativity, and sacrifice. However, until now, we have never had the leverage to change the equation. AI and emerging technologies may not erase our challenges, but they offer something EMS has rarely had: a force multiplier.

EMS has long been built around reaction—rising to meet each crisis as it comes. Our legacy systems reflect that history: designed for response, not resilience. But the future demands more. Today, we can reimagine EMS as a proactive, connected, and forward-thinking system that anticipates needs, strengthens coordination, and supports clinicians before burnout begins.

AI offers the architecture for that transition. It enables a shift from reactive care to anticipatory practice where EMS responds, predicts, prevents, and partners across the continuum of care. It reduces the daily friction that wears down both clinicians and agencies. It turns raw data into insight, insight into action, and action into meaningful, sustainable change.

Reflecting on the historical significance of EMS and the visionaries who shaped it, such as Dr. Peter Safar, we see a parallel in today's AI advancements. Just as Safar envisioned a national EMS information system in 1969, the integration of AI represents the next leap forward, offering a modernized, efficient, and responsive EMS system. Embracing AI is not just an opportunity but a necessary evolution to ensure that EMS continues to serve as the vital safety net of our society and healthcare system.

We have already seen this transformation in hospitals, health systems, businesses, and government. When deployed thoughtfully, AI has changed what is possible. It cannot solve every problem, but it can stabilize the foundation. It can allow us to move from crisis response to forward momentum. It will not fix everything but might empower us to fix what matters most.

When EMS intentionally applies this technology, the vision is no longer abstract but transformational. AI-powered documentation offers the potential to reduce burnout and liability, while real-time decision support can enhance accuracy during critical events. Furthermore, integrating field data into the patient's permanent record strengthens clinical continuity and closes gaps in care. We can show our value—not theoretically, but measurably, in real-time. No more documentation that disappears into silos. No more assessments that never reach the care team. What we see at the point of care—on scene, at home, or in transit—can become the cornerstone of what comes next.

In rural and frontier communities, autonomous vehicles may expand transport capacity for low-acuity or interfacility transfers. More innovative dispatch systems can better match clinicians to patient needs and stop sending lights-and-sirens responses to calls where speed makes no difference, but risk does. Predictive scheduling models can anticipate demand and deploy teams with precision. AI allows us to build more intelligent systems without losing the human connection that defines EMS.

This is not science fiction. It is what happens when the right tools meet the right people, focused on the right goals. The next chapter of EMS has not yet been written. But for the first time in a generation, we are not just reacting to the future—we have the power to help shape it. With the right investments and innovations, EMS can lead a new era of excellence in emergency medical care.

Let's get started, because leading EMS into the future will not happen by accident. It will happen because we choose to act.

Key Terms and Definitions

- **Artificial Intelligence (AI):** A category of computer systems designed to perform tasks that typically require human intelligence, such as recognizing patterns, making predictions, understanding language, and drawing conclusions based on probability rather than certainty.

- **Augmented Reality:** An interactive technology that overlays digital information—images, instructions, or alerts—onto the real-world environment. Used in EMS training, clinical decision support, and remote medical guidance.
- **Burnout:** A physical and emotional exhaustion caused by prolonged stress and overwork. It is often seen in high-demand EMS settings and positioned in the chapter as a challenge that can be mitigated by technology-driven solutions that reduce administrative and operational burdens.
- **Emerging Technology:** Any tool or system actively transforming health care in adjacent fields (e.g., hospitals, public health) but not yet widely adopted in EMS. These technologies are not experimental—they are in motion, tested, scaled, and refined.
- **Force Multiplier:** A tool, strategy, or intervention significantly enhancing impact, productivity, or efficiency. In the EMS context, AI and automation are force multipliers that amplify clinician capacity, reduce errors, and improve care delivery.
- **Fragmentation:** A structural characteristic of EMS in which policies, standards, oversight, and systems vary significantly across jurisdictions. Often linked to the profession's decentralized origins and cited as a persistent barrier to unified innovation and sustainability.
- **Natural Language Processing:** An AI capability that enables computers to interpret, process, and generate human language. In EMS, natural language processing supports real-time documentation, translation, and communication.
- **Predictive Analytics:** AI tools that use current and historical data to forecast future clinical or operational outcomes. In EMS, this may include identifying high-risk patients, anticipating call volumes, or preventing adverse events.

Reflection Questions

1. Reflect on your agency's current cultural attitude toward innovation. What historical factors or past experiences may have influenced this attitude? How can you overcome resistance to technology adoption?
2. Evaluate the current systems in place at your agency. Where do you see emerging technologies, such as AI or predictive analytics, making a significant impact? What are the potential obstacles to implementation, and how can they be overcome?

3. Imagine your agency could demonstrate its value to the broader healthcare system using technology. What data or outcomes would you want to showcase? How could technology help tell that story more effectively?
4. Think about a recent challenge your agency or system faced—whether related to staffing, documentation, or decision-making. Could any of the technologies discussed in this chapter have provided a solution? How would it have changed the outcome?
5. Consider the role of EMS in the future healthcare landscape. What would it look like for EMS to lead the next phase of healthcare transformation? What steps can you take today to help your agency contribute to this leadership?

Study Questions

1. What major federal policy change in the early 1980s led to the end of targeted national funding and leadership for EMS system development?
2. Which of the following best describes the early structure of EMS after federal funding was eliminated?

    a. Federally standardized
    b. Hyperlocal and community-driven
    c. Private and for-profit
    d. Run entirely by hospitals

3. EMS was developed alongside nursing and medicine as a fully supported federal healthcare profession from its inception.
4. True or False: EMS must shift from being a _______ system to becoming a proactive, predictive, and resilient system.
5. What are three core challenges currently threatening the sustainability of EMS, as described in chapter 1?
6. Several states have launched task forces or formal studies to evaluate EMS sustainability. What common themes did these initiatives identify as threats to the future of EMS systems?
7. Why does the chapter argue that AI should be embraced in EMS leadership?
8. What distinguishes meaningful EMS innovation from superficial upgrades?
9. According to the chapter, what is the risk EMS faces if it fails to lead or adapt to emerging technology?

    a. Decreased patient acuity
    b. Lower training standards
    c. Becoming irrelevant in a transforming healthcare system
    d. Increased vehicle maintenance costs

10. Why is understanding the legacy and cultural roots of EMS resistance to innovation important for leading future change?

CHAPTER SUMMARY

Innovation, Intelligence, and the Identity Crisis in Emergency Medical Services

EMS is at a critical inflection point, where the long-standing pressures of workforce shortages, financial instability, fragmented systems, and professional invisibility collide with a rapidly advancing healthcare landscape. Rooted in a history of federal disinvestment and decentralized development, EMS has inherited a culture of localism and skepticism toward large-scale innovation. That cultural legacy still shapes how many professionals respond to new technology, policy, and change.

Yet change is no longer optional. AI and related technologies are already transforming hospitals, dispatch centers, and public health systems. Like all healthcare sectors, EMS is undergoing a significant change. By strategically embracing AI, we can navigate these changes and emerge stronger and more effective than ever. This chapter positions AI not as a threat, but as a force multiplier: a set of tools that can help address chronic challenges like burnout, documentation overload, and system inefficiency. When used wisely, AI and automation can strengthen clinical decision-making, integrate EMS data into the broader healthcare continuum, and elevate the profession's visibility and impact.

The chapter also calls attention to the distinction between meaningful innovation and superficial upgrades. In a field saturated with friction and under-resourced systems, the focus must remain on tools that reduce burden, support clinicians, and improve care delivery. This is not about chasing trends. It is about defining what matters and ensuring EMS helps lead, not just follow, the transformation ahead.

2

Foundations

Artificial Intelligence

LEARNING OBJECTIVES

After reading this chapter, the reader will be able to

- Define AI and its core subfields—including machine learning, natural language processing, computer vision, and robotics—and explain their relevance to ems operations.
- Describe the historical development of AI in health care, identifying key milestones such as the MYCIN project, neural network breakthroughs, and the ImageNet competition.
- Analyze current use cases of AI in ems, including AI-powered documentation, predictive dispatch, real-time clinical decision support, and autonomous logistics.
- Evaluate the practical and ethical risks of AI implementation in ems, including algorithmic bias, hallucination, technical debt, data quality issues, and clinician accountability.
- Apply the technology adoption curve framework to assess EMS's position relative to other healthcare sectors in AI readiness and integration.
- Distinguish between technological novelty and meaningful innovation, identifying which AI tools function as true force multipliers for clinical performance, sustainability, and system design.
- Develop a leadership mindset toward AI integration, including strategies for evaluating tools, managing risks, preparing infrastructure, and ensuring responsible, equitable deployment.

Not Science Fiction—Just Science, Finally Reaching Emergency Medical Services

You did not enter the EMS field to become a software programmer and are not here to build algorithms or decode tech jargon. You are here to care for patients, your team, and the communities you serve. But as the healthcare system evolves around us, having a basic understanding of AI is no longer optional. It is essential.

AI is not merely changing health care; it's detonating its economic landscape. According to the SkyQuest Technology Group, by 2031, AI in health care is projected to command a global market of $286.77 billion (USD), with a staggering 38% compound annual growth rate.[1] That's explosive. Across all sectors, forecasts predict AI will exceed $1.3 trillion by 2030.

This chapter will not ask you to code. It will not turn you into a data scientist. Instead, it will equip you with a practical understanding of what AI is, how it works, and why it matters—right now—for EMS.

Hospitals, insurers, clinics, and health systems are already deploying AI to document faster, triage more accurately, and predict patient deterioration before a clinician enters the room. In some systems, that shift has already redefined how care begins—and who delivers it. EMS cannot afford to remain on the sidelines.

AI is not magic. It is not science fiction. It is science that finally works—and it is coming to the field. Whether EMS leads that change or is reshaped by it will depend on the decisions we make today.

The Core Tools of Artificial Intelligence—What Emergency Medical Services Leaders Need to Know

AI is not a single product, platform, or system. It is a category of tools—each designed to simulate specific aspects of human intelligence, such as learning, pattern recognition, language interpretation, and decision-making. For EMS leaders, understanding these tools does not require coding experience. It requires clinical insight and operational awareness. In other words, you do not need to build the future of AI, but you must be ready to lead EMS through it.

Machine learning (ML) is the most widely used form of AI in health care today. These systems learn from large volumes of data, detect patterns, and improve their performance over time. In EMS, ML is already being applied to predict patient deterioration, flag abnormal electrocardiograms (ECGs) or vital signs before a medic sees them, and assist dispatch centers by identifying high-risk calls based on caller language, computer-aided design history, and location. Unlike traditional software, ML does not rely on rigid programming. It adapts based on what it observes, continuously refining its outputs. For EMS clinicians and leaders, this means faster, more informed decision-making grounded in statistical probability rather than static rules.

Natural language processing (NLP) allows computers to understand, generate, and respond to human language. In the prehospital environment, NLP transforms how clinicians document care and communicate across language barriers. These tools can transcribe voice input into electronic patient care reporting (ePCR) narratives, extract relevant details from chaotic field environments, and even translate patient speech in real time, ensuring that critical clinical information is not lost due to language or stress. For many EMS agencies, documentation is one of clinicians' most burdensome and error-prone tasks. NLP offers a way to reduce that burden without compromising accuracy. It enables clinicians to spend less time typing and more time engaging with patients, partners, and care teams.

Yet it is essential to remember what AI is not. It is not sentient. It does not possess judgment, emotion, or awareness. It does not "know" in the way a human clinician knows. AI systems operate through mathematical models—processing inputs, identifying statistical relationships, and producing outputs. They are fast, but not wise. That distinction matters. Poorly implemented AI can reinforce bias, magnify bad habits, or automate inefficiencies. But when carefully designed and appropriately supervised, these tools can increase diagnostic accuracy, relieve mental workload, highlight risks that would otherwise be missed, and help clinicians refocus on what matters most: patient care.

Computer vision is another emerging application that is beginning to shape EMS operations. This form of AI enables machines to interpret visual information, such as photos, video, or real-time camera feeds. In EMS, computer vision supports injury recognition from body-worn cameras, monitors ambulance driver behavior to detect distraction or fatigue, and enhances telemedicine platforms by interpreting facial cues and movement during remote triage. These tools do not replace human observation. They supplement it, adding layers of analysis that can improve safety, accountability, and clinical decision-making in high-stakes environments. Computer vision can also "read" prescription medication lists, hospital facesheets, and handwritten notes—helping clinicians quickly extract key information in the field when time, lighting, or cognitive load make manual review difficult.

Robotics and drones are expanding what EMS systems can do, especially in geographically isolated or resource-constrained environments. While the term "robotics" may conjure images of humanoid machines, the reality in EMS is more practical. Drones are already delivering automated external defibrillators, naloxone, blood products, and medications to scenes where response times are otherwise too long. In some communities, autonomous vehicle pilot projects are being tested for medical transports and low-acuity patient transfers. In disaster zones, robotic tools operate in environments that are too dangerous for human responders. These technologies do not simply add capacity—they extend it, allowing EMS to operate more efficiently, safely, and with greater reach than ever before.

Individually, each of these tools offers an incremental benefit. But together, when ML supports risk stratification, NLP automates documentation, and drones accelerate logistics, they become something more. They become the foundation for an intelligent, integrated EMS system. The challenge ahead is not just about adoption. It is about leadership—ensuring these tools are used wisely, ethically, and in service to both clinicians and the patients they care for (fig. 2–1).

What Is a Perceptron?

A **perceptron** is the simplest type of artificial neuron—the building block of many AI systems in health care today. It takes multiple inputs (like symptoms or vital signs), assigns them importance (weights), sums them, and passes the result through a threshold to produce an output (e.g., "at risk" or "not at risk"). This basic function allows the system to classify information and learn from mistakes. Modern EMS tools that flag ST-elevation myocardial infarctions or identify abnormal patterns often trace their lineage back to this 1957 innovation.[9]

A Short History of a Long Evolution

AI may feel like a sudden arrival, but its roots stretch back nearly seventy years. Long before EMS became a profession, the notion of machines mimicking human thought took hold. In 1956, at a pivotal Dartmouth College gathering, scientists asked a groundbreaking question: Could machines truly reason, learn, and decide? From that meeting, the term "artificial intelligence" emerged, launching decades of research—a journey filled with both triumph and setback.[2]

The earliest systems held promise, but the technology simply was not ready. One of the first significant healthcare efforts came in the 1970s with *MYCIN*,[3] an AI program developed at Stanford University to recommend antibiotic treatments. MYCIN often outperformed physicians[4] in clinical accuracy but it never reached the bedside—not because it failed scientifically, but because it lacked

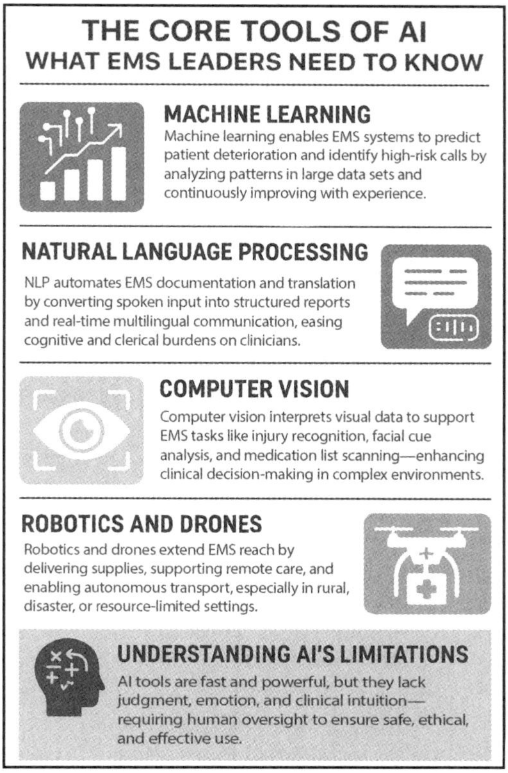

FIGURE 2–1. Understanding the foundational tools of AI—ML, NLP, computer vision, and robotics—is essential for EMS leaders preparing their agencies for digital transformation. Human judgment must remain as the anchor that keeps the new AI tools safe and ethical when used in patient care.

integration, liability protection, and clinical trust. It was ahead of its time and proof that innovation without usability goes nowhere.

Progress accelerated through the 1980s and 1990s. Researchers developed neural networks that mimicked the structure of the human brain, allowing computers to "learn" from their mistakes. Convolutional neural networks were introduced, enabling image recognition that would later be used to analyze ECGs, chest X-rays, and video feeds. The breakthroughs were technical, but their real-world impact was limited by computing power, data access, and public trust.

Then came 2012. In a global competition called the ImageNet Challenge, a deep learning model known as *AlexNet* shattered records for computer vision.[5] It recognized images with unprecedented accuracy—and did so faster, better, and more efficiently than any system before. That moment ignited the modern AI revolution. Hospitals began using AI to read radiology films, detect sepsis earlier, and predict which patients were at risk of crashing. Electronic health record vendors started embedding AI into clinical workflows. Insurance companies created algorithms to detect fraud and flag high-risk claims. Technology was no longer theoretical. It was operational.

But EMS was not part of that wave. While the rest of health care embraced AI, EMS lagged, not because of a lack of need but because of fragmentation, underfunding, and outdated procurement models. Our systems were built to respond, not to innovate. As hospitals deployed real-time analytics, we struggled with incompatible software, legacy ePCRs, and paper protocols.

How Artificial Intelligence Learns from Mistakes: A Simple Analogy

Backpropagation is the process AI uses to improve its accuracy. Think of it like teaching a child to throw a ball into a basket. The child throws, misses, and you offer guidance—"aim higher" or "throw softer." With each attempt and bit of feedback, the child adjusts. Neural networks do the same thing. They make a prediction, compare it to the correct answer, calculate the error, and work backwards through their layers to adjust internal settings. Repeating this process allows the system to "learn" from its mistakes and get more accurate over time. It is not magic. It is just feedback at machine speed.

Today, that gap is no longer just inconvenient—it is dangerous. EMS can no longer afford to operate as if innovation can be delayed for years or entirely ignored. The breakthroughs that reshaped emergency departments and intensive care units are now accessible to the field. The technology works. The question is whether EMS leaders will help shape its use or inherit a system designed without us.

AI used to be an academic exercise—researched in universities, modeled in code, and discussed in theory. Today, it is embedded in electronic health records, billing systems, telehealth platforms, and mobile apps. The shift is no longer technical—it is operational. And EMS is next. The question is not whether we will engage. It is about whether we will shape how that engagement happens, or be handed systems we did not help design.

Milestones That Made Artificial Intelligence Work—Why Emergency Medical Services Should Care

1957: The Perceptron
Dr. Frank Rosenblatt developed the first artificial neural network. It mimicked how the brain processed input—planting the seed for today's predictive algorithms.

1970s–1990s: AI in Medicine
Stanford's MYCIN system outperformed physicians in recommending antibiotics, but never reached patients due to workflow and legal concerns—a warning about deploying unintegrated tech.

1980s–1990s: Neural Networks
Researchers like Hinton and LeCun built convolutional networks and refined backpropagation, laying the foundation for computer vision and natural language understanding.

2012: AlexNet Breakthrough
A deep learning model called AlexNet outperformed every competitor in the ImageNet Challenge. It sparked the AI revolution now transforming health care—everywhere except EMS.

Despite its promise, AI was not built with EMS in mind. When MYCIN emerged in the 1970s, EMS was still in the early stages of defining a paramedic. Likewise, when AlexNet ignited the modern AI revolution in 2012, many EMS agencies struggled with converting from paper patient care reports to ePCRs. These breakthroughs happened without our input because EMS lacked the funding, research infrastructure, and policy visibility to participate. AI matured in other branches of medicine while EMS remained isolated. This is not a historical footnote—it is a warning. If EMS is left out of the next wave of AI advancement, it will not be because the tools were unavailable. It will be because our systems were unprepared to adopt them.

The Tools Already Transforming Emergency Medical Services

AI is not on the horizon—it is already in the building. Across hospitals, clinics, and public health systems, AI-powered tools are actively redefining how health care is delivered, documented, and coordinated. EMS is no longer observing this change from a distance. These technologies are beginning to reach the field. The question is not whether they are coming but whether EMS leaders are ready to use them wisely.

Emerging technologies already influence EMS practice today. They are no longer speculative or experimental. They are deployed, tested, and—when implemented thoughtfully—deliver measurable impact. Each offers a window into what EMS could become when innovation meets clinical intent.

As the tools mature, the use cases become clearer. EMS does not need speculative promises—we need proven gains. The most promising applications directly address our real-world challenges: documentation fatigue, diagnostic uncertainty, and system inefficiency. AI's value proposition becomes much more tangible when evaluated through that lens.

AI-powered documentation is one of the most readily applicable AI tools in EMS. Voice-to-text systems, smart narrative generators, and context-aware prompts empower clinicians to complete patient care reports faster and more accurately. Some systems even auto-populate ePCRs from structured field data or spoken input, directly addressing after-call charting time and documentation fatigue. Hospitals have already integrated these tools into emergency department workflows, and EMS is just beginning to catch up.

AI-powered documentation should have already been implemented. EMS has lagged years behind hospitals in adopting these tools, even though field documentation is often more complex, variable, and time-constrained than inpatient charting. This delay is not due to a lack of value but to inertia, funding constraints, and vendor fragmentation. EMS deserves the same efficiency and protection these tools now provide elsewhere.

In May of 2024, Mediwave launched the first custom-built AI-powered ePCR platform designed specifically for EMS. The system uses NLP to transcribe real-time field narratives and auto-tags the corresponding fields in the National Emergency Medical Services Information System. It eliminates the need for dropdown menus, improves documentation accuracy, and significantly reduces time spent on post-call reporting. For many EMS agencies, especially those facing staffing shortages or high call volumes, tools like this offer a breakthrough, restoring time, reducing error, and protecting clinicians from administrative overload.

Predictive analytics and risk stratification tools help forecast which patients will likely deteriorate, require advanced life support transport, or need early hospital activation. These ML systems analyze historical and real-time data to guide dispatch prioritization, dynamic triage, and resource allocation. They do not replace clinical judgment but enhance it, particularly in time-critical or resource-limited scenarios.

Real-time clinical decision support is reshaping how care is delivered in the field. Unlike traditional protocols, which are static and linear, AI-enabled systems adapt to the patient's presentation. They interpret vital signs, scene details, and historical data to provide evidence-informed, real-time guidance. This support can improve accuracy, reduce errors, and strengthen confidence for clinicians facing high-acuity situations.

AI-enabled dispatch and call triage tools can now identify cardiac arrest, respiratory distress, and behavioral emergencies based on live speech analysis. These tools analyze caller tone, language, and background noise, sometimes detecting urgency before the words make it clear. In systems where every second counts, this early recognition can accelerate CPR initiation and reduce time to intervention.

In early 2024, Life EMS Ambulance[6] in Grand Rapids, Michigan, deployed a predictive dispatch platform that uses AI to forecast emergency call locations based on historical patterns. Predictive dynamic deployment systems analyze time of day, geography, and incident type to position ambulances proactively across the service area. This approach has reduced response times and improved unit availability without increasing staff. For EMS leaders, this case illustrates how AI can optimize operations, helping agencies do more with limited resources while improving patient access to timely care.

Workforce optimization and crew scheduling tools are also gaining traction. By using AI to analyze shift history, call volume, geographic patterns, and clinician availability, these platforms can recommend schedules that match the right team to the right call at the right time. Some systems can automatically identify the closest off-duty paramedic with the appropriate skill set and dispatch them as backup.

AI-powered billing and revenue cycle management is already improving agency sustainability. These platforms scan narratives, identify missing elements, flag incomplete records, and predict coding errors that might trigger denials. They protect revenue and compliance by streamlining claim submission and reducing rework.

These systems offer a glimpse of what longitudinal EMS care could look like. Imagine a future in which paramedics are alerted to changes in a high-risk patient's vitals before symptoms become visible. AI analyzes continuous monitoring data, flags deviations from baseline, and prompts a community paramedic check-in, preventing a full-blown emergency before 9-1-1 is ever called. This is not a distant possibility; it is the convergence of predictive analytics, wearable tech, and expanded EMS scope.

Financial sustainability is not just about billing faster—it is about capturing revenue that would otherwise be lost. In many EMS systems, incomplete documentation, missing signatures, or imprecise coding lead to denials, delays, and underpayments. AI can proactively identify and mitigate these risks before submission, increasing collections without adding staff or overtime. This is not a luxury for rural and underfunded agencies—it is survival.

Wearables and remote monitoring devices are enabling EMS to engage patients before the call and after the handoff. These tools transmit live biometric data to EMS teams or hospitals, offering early warning of clinical deterioration or confirmation of clinical stability. They are especially valuable in telehealth, community paramedicine, and chronic care models.

These systems offer a glimpse of what longitudinal EMS care could look like. Imagine a future in which paramedics are alerted to changes in a high-risk patient's vitals before symptoms become visible. AI analyzes continuous monitoring data, flags deviations from baseline, and prompts a community paramedic check-in, preventing a full-blown emergency before 9-1-1 is ever called. This is not a distant possibility; it is the convergence of predictive analytics, wearable tech, and expanded EMS scope.

Multilingual communication tools powered by AI are breaking language barriers that have long undermined patient safety. Real-time speech translation and multilingual ePCR interfaces enable clinicians to understand and document encounters with patients who speak little or no English. These tools improve equity and protect against clinical error in high-diversity settings.

AI can enhance and expand today's EMS clinicians into even stronger healthcare professionals when implemented purposefully. In rural and frontier areas, where backup may be hours away and resources are scarce, AI-powered tools can provide real-time insights, support critical decisions, and ensure that EMS clinicians are not operating alone. These technologies are not replacing field medicine; they are amplifying it, turning moments of isolation into opportunities for precision care.

Autonomous vehicles and drone-based logistics are no longer science fiction. In some systems, drones deliver naloxone, blood products, and trauma kits directly to the scene, sometimes before responders arrive. Driverless ambulances are being piloted for interfacility or low-acuity transport. These technologies are expanding geographic reach and transforming how logistics is managed.

Augmented reality and virtual reality are used in EMS training and clinical support. Augmented reality overlays instructions or alerts on the real-world environment, which is helpful for remote trauma guidance, airway management, or incident command. Virtual reality supports immersive simulation for mass casualty events, advanced airway procedures, and team-based decision-making. These tools are improving training and changing how EMS learns, rehearses, and prepares.

In Sri Lanka, this kind of forward-looking integration is already underway. The country's connected ambulance system merges augmented reality with

AI-powered clinical decision support, delivering real-time guidance and heads-up data overlays in the field. Through devices like the Microsoft HoloLens, emergency medical technicians receive procedural support, vital sign monitoring, and telemedicine consultations while maintaining focus on the patient. It is not a prototype—it is national infrastructure. This international example proves that intelligent, immersive technology is feasible and scalable for EMS systems in the United States.

None of these innovations requires the reinvention of EMS. They require intention—leadership that understands how technology can support clinicians, not distract from them. The tools already exist. The opportunity is to use them to build a smarter, safer, and more sustainable EMS system—before that opportunity passes us by.

Other healthcare professions have already embraced this shift. Nursing, in particular, has taken a proactive role in shaping how AI supports—not replaces—clinical care. In a 2022 position statement, the American Nurses Association emphasized that the "appropriate use of AI in nursing practice supports and enhances the core values and ethical obligations of the profession." They called on nurses to evaluate AI's impact on patient outcomes, educate patients and families, and influence how the technology is developed and deployed (fig. 2–2). This is not a passive endorsement. It is a model for how a profession can lead from within, ensuring that innovation aligns with practice, ethics, and patient trust. EMS must do the same.

Understanding the Risk and Leading Responsibly

AI offers robust solutions to some of EMS's most persistent challenges, but potential does not equal protection. These tools can streamline documentation, support clinical decisions, and optimize logistics—but only if deployed responsibly. If adopted unquestioningly, AI can magnify flaws, embed bias, and introduce new forms of harm.

The first and most widely misunderstood risk is *AI hallucination.* This occurs when an AI system generates inaccurate, misleading, or entirely fabricated content but presents it confidently. In clinical documentation, that kind of error is serious. It can result in documented false symptoms, incorrect narratives entered into medical records, or faulty data used in protocol reviews, quality assurance (QA) audits, or billing. These systems are designed to fill in the blanks based on patterns and probability, not certainty or context.[7] Without

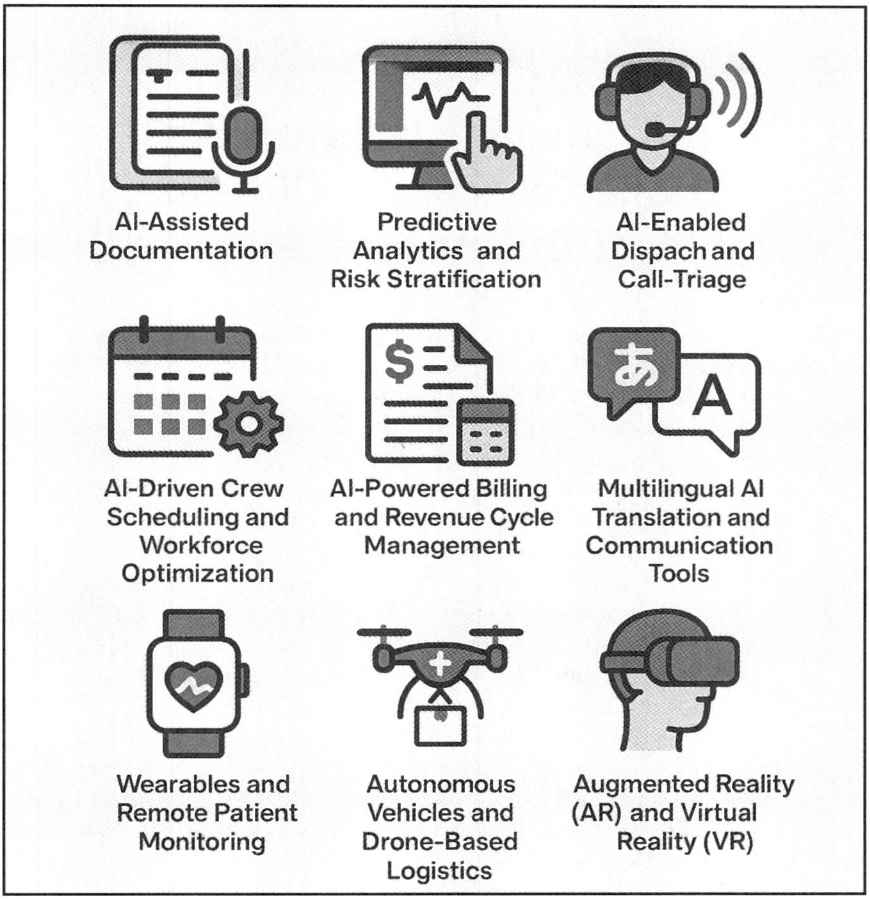

FIGURE 2–2. AI technologies are already influencing nearly every aspect of EMS—from predictive analytics and documentation to multilingual communication, logistics, and immersive training through augmented and virtual reality.

human review, a well-intentioned tool can quietly introduce dangerous inaccuracies into the clinical record.

Equally critical is the problem of insufficient data. AI learns from what it is given, and in EMS, that data is often inconsistent, incomplete, or inaccurate—including incomplete timestamps, erroneous click-box entries, or narratives written in haste after long shifts. If flawed data is used to train an AI model, the system will learn to reproduce and reinforce those errors. Instead of elevating care, it will automate dysfunction. Predictive triage models built on biased or messy data can lead to poor decision-making and the amplification of inequities, rather than solving them.

CASE IN POINT

Artificial Intelligence Hallucinations in Clinical Documentation

Imagine this scenario: An EMS agency deploys an AI-powered documentation assistant trained on thousands of past patient care reports. Over time, the system learns that chest pain patients with confirmed ST-elevation myocardial infarctions typically receive a 12-lead ECG, aspirin, and morphine. These patterns are statistically reinforced across countless records.

Now, you are a paramedic caring for a patient with chest pain and ECG changes, but you do not administer aspirin to the patient. Maybe the patient was allergic, or perhaps you were distracted and forgot. Either way, no aspirin was given. But unless you tell the AI model those details, the AI system is designed to complete the record based on probabilities, not real-time clinical judgment. So, it fills in the blank: "Aspirin 324 mg administered PO."

That is a *hallucination*—fabricated content that never happened, inserted confidently into a legal medical record. It creates a false narrative, misrepresents care, and introduces risk in quality assurance reviews, medical oversight, or even courtroom testimony.

AI tools must be programmed to flag missing or unusual data, not fabricate it. Every narrative still requires clinician review and accountability. The integrity of the patient record—and patient safety—depend on it.

That leads directly to the issue of algorithmic bias. AI systems reflect the data and the humans used to build them. If a model is trained mostly on urban calls, it may underperform in rural settings. If the data lacks demographic diversity, it may misclassify conditions in minority populations. In a profession where seconds matter and split-second decisions determine outcomes, this bias can result in under-triage, missed diagnoses, or inequitable care. These are not hypothetical concerns. Facial recognition systems have been shown to underperform for individuals with darker skin tones.[8] Clinical risk scores have deprioritized care based on flawed racial proxies. Dispatch algorithms have misread urgency based on accents or language differences. The technology itself is not inherently biased, but if EMS leaders do not demand transparency, diversity in QA, and explainability in AI outputs, we risk embedding systemic inequities into our digital infrastructure.

For EMS, the risk of bias is operational. If an AI-enabled dispatch tool fails to recognize nonstandard speech patterns or underrepresents rural geographies in its training data, lives could be lost. Likewise, clinical decision support tools must be evaluated across diverse patient populations and validated in the unique context of prehospital care. Without deliberate QA and demographic analysis, AI could replicate existing disparities—or create new ones.

Another overlooked risk is technical debt. Most EMS systems operate on aging software and legacy infrastructure. Introducing AI into that ecosystem, without modern integration, can cause severe disruption: delays in call entry, crashes during documentation, or double work due to incompatible systems. Some vendors will promise "plug and play" solutions, but the reality is rarely that simple. Without proper IT support, test environments, and staged deployment, AI can create more problems than it solves.

There are also critical privacy, security, and liability questions. AI systems often process enormous volumes of patient data, much of it in real time. Agencies face legal exposure if that data is mishandled through unencrypted application programming interfaces, unsecured cloud platforms, or third-party vendors. The Health Insurance Portability and Accountability Act (also known as HIPAA) violations, patient harm, and organizational liability are not abstract risks. EMS leaders must demand strong business associate agreements, confirm encryption standards, and ask hard questions about data ownership, audit trails, and legal accountability. If a system makes a recommendation that causes harm, who is responsible? If an AI-generated narrative contains false data, who defends it in court?

Finally, EMS must address the ethical blind spots AI introduces. As these systems influence triage, protocol adherence, or transport decisions, clinicians may feel pressure to follow a machine's guidance—even when it contradicts clinical intuition. This creates a dangerous ambiguity around accountability. Who explains the decision when AI down-triages a critical patient? Who answers to the family when a system error results in harm? EMS must continue to consist of human override points, ethical review processes, and education that reinforces the clinician's foundational role as the final authority on patient care.

These risks are not reasons to reject AI. There are reasons to lead its integration deliberately, transparently, and ethically. What is at stake is not just operational efficiency—it is trust. Trust in our systems, trust in our documentation, and trust in our clinical decisions. If we handle this transition well, AI becomes a powerful partner. If we do not, EMS becomes a cautionary tale.

Many challenges—data quality, algorithmic bias, documentation integrity, and ethical accountability—will be explored in greater depth in chapter 11. That chapter addresses the real-world stakes of integrating AI into prehospital care, including clinician responsibility, patient equity, legal oversight, and the systems-level governance needed to support ethical innovation. For now, EMS leaders must begin with awareness: We cannot afford to adopt technology that moves faster than our ethics. Every AI deployment must have safeguards, clinical oversight, and a clear understanding that the clinician, not the algorithm, is ultimately accountable (fig. 2–3).

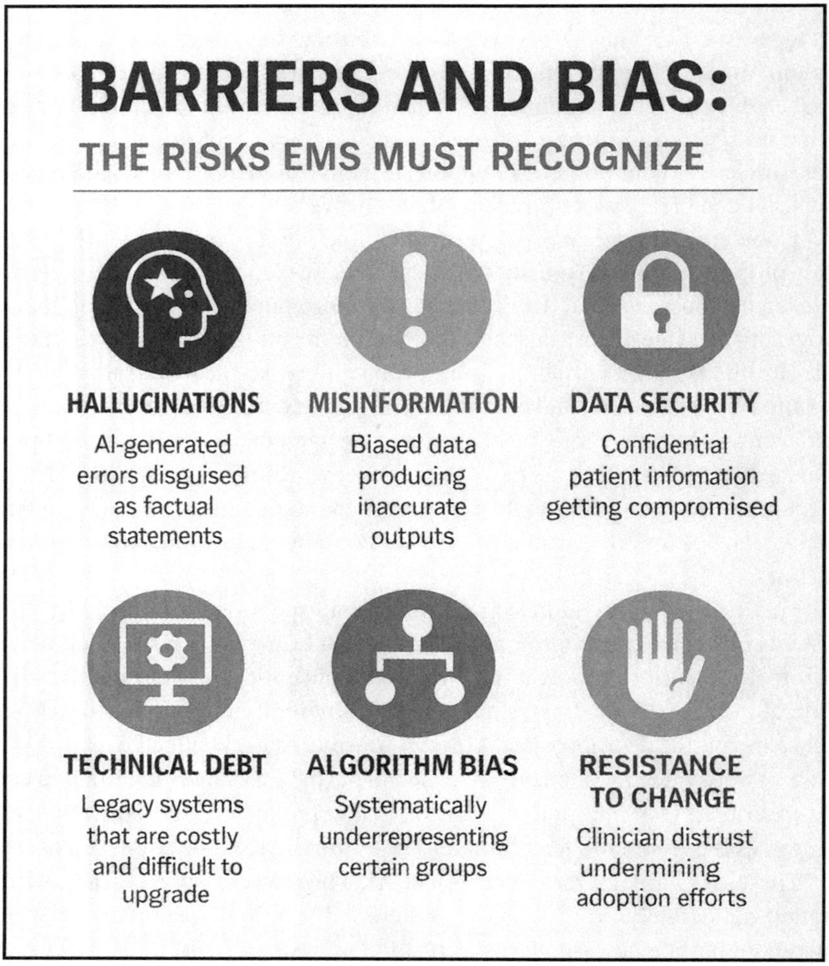

FIGURE 2–3. AI carries both promise and peril. Hallucinations, bias, and data security risks demand vigilant oversight.

The Adoption Curve and Where Emergency Medical Services Are Falling Behind

Innovation's spread is rarely instantaneous; it unfolds gradually and often unevenly. Everett Rogers, a sociologist, illustrated this process in 1962 with the technology adoption curve, which categorizes adopters into five groups: innovators, early adopters, early majority, late majority, and laggards (fig. 2–4).

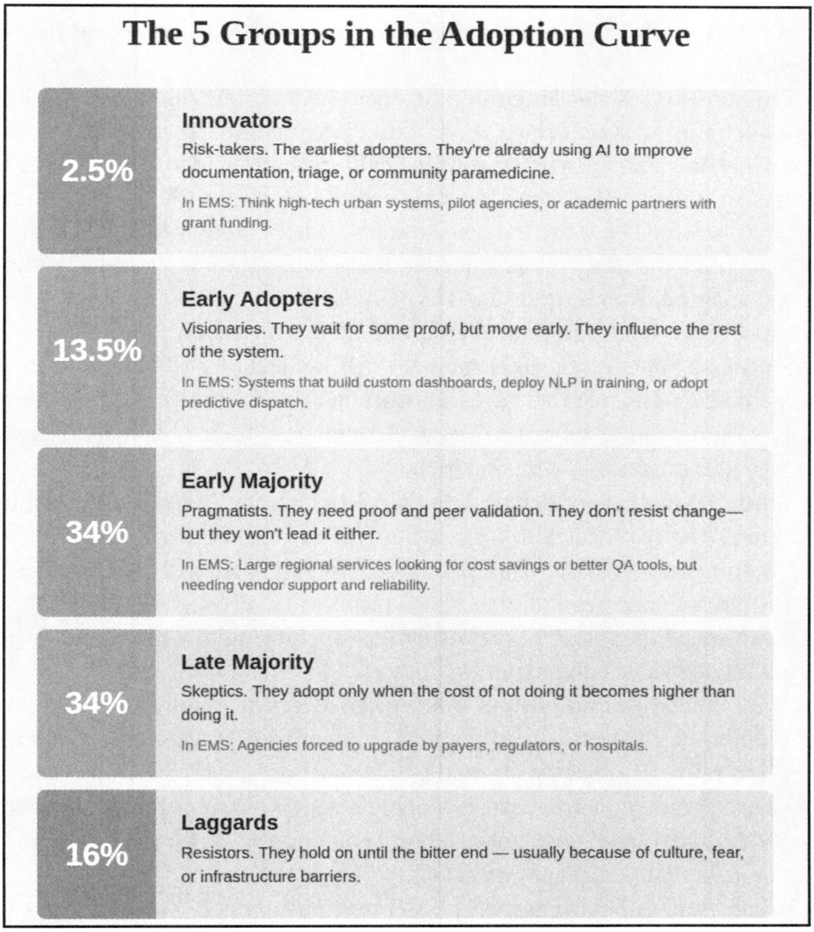

FIGURE 2–4. Every innovation follows a predictable path. From risk-taking innovators to cautious laggards, recognizing where your agency falls on the adoption curve helps leaders plan realistic AI implementation strategies.

Though the framework originated in agriculture and consumer products, its relevance to health care, particularly EMS, is hard to ignore.

Across the curve, each group reacts differently to change. *Innovators* are risk-takers, quick to experiment, and willing to fail fast. *Early adopters* follow closely, often driving momentum. The *early majority* waits for proof. The *late majority* only moves when not changing becomes more painful than adapting. And *laggards* resist until change is unavoidable—or imposed.

In the broader healthcare system, AI is steadily moving from the early adopter to the early majority phase. Hospitals use predictive analytics to manage bed flow, insurers flag claims with algorithmic review, and pharmacy chains triage minor conditions without clinician involvement. EMS, by contrast, sits further back—often in the late majority, and many systems are still among the laggards.

This position is not about competence. EMS clinicians are innovative, adaptable, and mission-driven. EMS clinicians operate in some of the most cognitively demanding environments in health care, often managing high-acuity patients in uncontrolled, low-resource conditions with little support and no margin for error. These are the environments where thoughtfully designed AI tools can offer the most value. But first, those tools must be built for the field, not just adapted from hospital models.

Unfortunately, structural and cultural barriers have slowed progress. Most EMS agencies operate on tight budgets, rely on legacy software, and juggle fragmented governance structures. Procurement cycles are slow, integration is difficult, and cultural mistrust of new technology—fueled by years of unfulfilled vendor promises—remains high.

While EMS leaders debate whether AI-generated narratives constitute "cheating," the rest of health care is integrating real-time triage models and auto-documentation as standard practice. This gap is not just philosophical; it has operational and reputational consequences. By proactively adopting AI, EMS can shape the future of prehospital care, ensuring that new technologies align with our values and expertise.

The implications go beyond innovation, affecting funding, partnerships, and credibility. Insurers already use AI to identify high-value care pathways and align reimbursement with quality outcomes. If EMS does not integrate with these evolving metrics, we risk being categorized as outdated, low-value responders. That label does not just hurt reputations—it threatens reimbursement models that keep agencies afloat.

In my professional judgment, EMS currently hovers between the early majority and late majority phases on the technology adoption curve, well behind hospitals, payers, and some retail health systems. This lag is not just symbolic; it has tangible consequences. Agencies that fall behind miss out on innovation funding, get excluded from early pilot programs, and lose leverage in negotiating with vendors. The further EMS drifts from the front of the curve, the more likely others—hospitals, insurers, and tech companies—will define the profession's digital future. To regain leadership, EMS must reposition itself as an early adopter. That means more than purchasing new tools; it requires embedding innovation into agency culture, clinical practice, and system strategy.

Delaying innovation does not preserve autonomy. It increases the likelihood that EMS will be handed systems we did not shape, powered by models we did not validate, and governed by standards we did not write.

To move forward, EMS must deliberately shift its position on the curve. EMS can also benefit by borrowing strategies from healthcare sectors that have successfully scaled AI. Health systems that have succeeded did not begin by building AI from scratch—they started by identifying high-friction workflows and inviting clinicians to test simple, targeted tools that offered immediate value. EMS should follow the same path. Start where the pain is most significant—charting delays, staffing misalignment, delayed clinical recognition—and design pilots that relieve that pain first.

That means developing internal champions who understand the tools, identifying pilot projects that prove value, and demanding that vendors build systems for EMS, not retrofitted tools designed for hospitals. It means building tech literacy into leadership development, aligning policy with innovation, and creating feedback loops that include clinicians and patients.

This is not just about keeping up. It is about staying relevant. In a healthcare environment that rewards systems for efficiency, precision, and data-driven performance, EMS cannot afford to operate as if innovation is optional.

The adoption curve is not a judgment. It is a mirror. And what it shows us is clear: EMS is not out of time yet, but if we don't make changes soon, we may not be able to catch up.

For a deeper analysis of why EMS systems often struggle to adopt new tools—and what can be done to change that—see chapter 4, which explores the cultural, regulatory, and cognitive forces that shape our resistance to innovation. It also offers real-world examples and practical strategies for moving from inertia to momentum.

Delaying adoption does not protect the EMS profession—it places it at risk of losing control over its future. If EMS leaders do not shape how AI is integrated into dispatch, documentation, triage, and QA, others will. Hospital systems will design the tools, and expect EMS to conform. Payers will fund AI-optimized platforms—and penalize those that are not. Tech companies will build systems around patients, not clinicians. These shifts are already underway. The longer EMS waits to engage, the more it will be governed by systems it did not choose.

When EMS leaders discuss the future, the conversation often centers on reimbursement, staffing, or scope. But the real question is whether we will shape the technology that is already shaping us. AI is not waiting for consensus. It is moving forward—training on our data, influencing our workflows, and transforming adjacent systems. Others will define it for us if we do not lead this change.

The Emergency Medical Services Dilemma: Stuck in Tradition, Starved for Innovation

EMS did not fall behind on purpose, but that is what has happened. While AI reshapes hospitals, urgent care centers, and public health infrastructure, most EMS systems remain stuck, not because of a lack of intelligence or courage but because of constraints calcified over time: funding, staffing, and imagination. If these are not confronted directly, EMS will not be disrupted from the outside; it will erode from within.

The EMS system was designed to respond, not to evolve. Its core metrics—response times, transport rates, cost-per-call—all reward volume over value. Most agencies operate without research and development budgets, without innovation officers, and without designated staff to test new technologies or evaluate emerging tools. Instead, they manage broken ePCR contracts, chronically understaffed QA teams, and protocol manuals that may not reflect the realities of modern care. Training programs, if they exist, are often squeezed into overtime hours—if there is any funding left at all.

This is not a failure of leadership. It is a failure of design. But here is the uncomfortable truth: No one else will fix it for us. Even the vendor ecosystem—those who claim to serve EMS—has often failed to deliver. Software companies frequently retrofit hospital tools and market them to EMS, regardless of operational fit. Dispatch systems lag in NLP. ePCR platforms lack real-time AI integration. Products are sold as "AI-enabled" when, in reality, they offer little more than automation. What EMS receives is often yesterday's technology, rebranded for today's crisis.

Meanwhile, the workforce is fraying. Clinicians are burning out under staffing shortages, documentation overload, and mounting pressure to do more with less. Ironically, many technologies that could reduce workload—AI-powered documentation, predictive staffing, real-time support—remain out of reach due to budget constraints or policy delays. EMS does not lack access to innovation; it lacks permission to use it.

This is how professions decline—not through one catastrophic failure, but through a steady erosion of capacity and relevance. While other healthcare sectors build future-ready systems, EMS risks becoming a legacy service—a field defined by what it *used* to do, not what it *can* become.

As an EMS leader, I believe our greatest existential threats are not theoretical. They are burnout, documentation overload, and the slow collapse of EMS in rural communities. These are not just symptoms of underfunding—they are warning signs of a system that cannot modernize fast enough to survive. The urgency is not abstract when our workforce is exhausted, charting takes longer than transport, and whole regions no longer have a reliable 9-1-1 response. It is immediate. AI will not fix these problems by itself, but it offers one of the few tools we have left to buy time, restore function, and begin to rebuild.

But that outcome is not inevitable. The dilemma we face is also an opportunity.

EMS does not need to match the scale or infrastructure of hospital AI deployments. We do not need billion-dollar investments or sprawling data teams. We need clarity, leaders who understand the tools, clinicians trained to use them ethically, and systems willing to pilot first and perfect later. Innovation will not wait for us to be ready, and patients will not wait either.

We have seen this lesson play out in other industries. Taxi companies did not lose business to Uber and Lyft because they were worse at driving—they lost because they ignored what was coming. By the time they responded, someone else had redefined the market. If EMS fails to integrate AI and emerging technology, we risk the same fate, not from malicious intent but misplaced complacency. Innovation does not care who got there first. It moves where there is vision, investment, and the courage to adapt.

This is not about chasing technology for its own sake. It is about rebuilding the foundation of EMS in a way that is smarter, stronger, and more sustainable. The profession is at a turning point—not because of AI itself, but because of what it forces us to confront: the fact that our current model, while courageous and committed, is no longer equipped to meet the demands of modern medicine.

Our mission is not to chase every innovation, but to ensure that the tools we adopt serve the patient, support the clinician, and sustain the system. That requires judgment, not hype. It requires leadership, not just enthusiasm.

We are not being left behind. We are choosing whether to step forward.

EMS has always adapted in adversity, but adaptation is no longer enough. The pace of technological change demands transformation. If we want to preserve EMS's mission, we must be willing to change its methods. That means investing in tools that reduce administrative burden, advocating for systems that reflect real-world operations, and preparing the next generation of clinicians to lead—not just follow—the evolution of care. This is not a luxury. It is a necessity.

EMS has constantly innovated under pressure, finding ways to adapt without funding, infrastructure, or institutional support; however, that same ingenuity can become a trap if we mistake workarounds for sustainability. AI offers EMS a rare opportunity not just to patch the system but to reimagine it.

Key Terms and Definitions

- **AI–Enabled Dispatch:** The use of AI to analyze call data and identify high-risk conditions, such as cardiac arrest or stroke, earlier and more accurately than human call-takers alone.
- **Autonomous Systems:** Technologies capable of operating independently, such as driverless vehicles and drones. In EMS, these systems are used for patient transport, logistical support, or delivery of life-saving supplies.
- **Clinical Decision Support:** Software or algorithms that provide real-time guidance to clinicians by analyzing available data and offering recommendations, alerts, or decision prompts based on the patient's presentation.
- **Documentation Burden:** The time, effort, and mental workload of recording patient encounters. This includes ePCR completion, narrative writing, and compliance documentation in EMS, often exacerbating clinician fatigue and liability risk.
- **Generative AI:** A type of AI capable of creating original content, such as text, images, or speech, based on user input. In EMS, this includes tools that generate ePCR narratives or voice-to-text documentation from structured prompts.
- **Machine Learning (ML):** A subset of AI that enables computer systems to identify patterns and improve performance over time based on data exposure. In EMS, ML powers predictive tools for triage, risk stratification, and resource allocation.

Reflection Questions

1. Which AI tools described in this chapter—ML, NLP, computer vision, or robotics—would benefit your EMS system today? What would it take to pilot it safely?

2. Does your current system have the infrastructure to support AI integration? If not, what's the most significant technical, financial, or cultural barrier?
3. How confident are you in the quality and integrity of your agency's data? Would you trust an algorithm trained on your system's ePCRs?
4. Where is your agency on the adoption curve? Are you leading, watching, resisting, or unaware of the curve?
5. What would it look like for your EMS system to proactively lead AI adoption rather than wait to react? What's one step you could take this quarter to move forward?

Study Questions

1. What is AI, and why is it critical for EMS clinicians and leaders to understand it today?
2. Which of the following is NOT a core AI capability discussed in the chapter?

   a. Machine learning
   b. Natural language processing (NLP)
   c. Virtual reality gaming
   d. Computer vision

3. True / False: AI is capable of sentient decision-making and emotional judgment in clinical situations.
4. In EMS, NLP can transcribe _______ into ePCR narratives and extract relevant clinical information.
5. What critical event in 2012 accelerated modern AI development and demonstrated practical use of deep learning?
6. Describe one real-world EMS application of predictive analytics or real-time decision support discussed in this chapter.
7. According to the technology adoption curve, what phase is EMS currently in when it comes to adopting AI?
8. List two major risks associated with poorly implemented AI in EMS.
9. Why is algorithmic bias a critical concern for EMS systems using AI-powered tools?
10. What lesson does the chapter draw from the example of taxi companies and Uber in the context of EMS and innovation?

CHAPTER SUMMARY

Foundations of Artificial Intelligence in Emergency Medical Services

AI is no longer theoretical—it's here and operational. However, most EMS systems remain reactive, navigating legacy systems while the rest of health care moves forward. This chapter clarified what EMS leaders must understand: AI is not a monolith. It's a toolbox, and those tools—ML, NLP, computer vision, and robotics—are already improving accuracy, documentation, logistics, and triage in other sectors.

We reviewed the history of AI's development and highlighted the critical breakthroughs that make it viable for field deployment today. We also addressed the real risks: hallucinations, insufficient data, bias, privacy failures, technical debt, and the ethical minefields EMS cannot ignore.

Then we turned inward. EMS is lagging not because of a lack of interest, but because of structural, cultural, and economic barriers. Most systems weren't built to innovate—they were built to respond. But that design is now a liability.

The technology adoption curve makes clear what's at stake. EMS is teetering between early adopter and early majority status, while health care at large races ahead. This delay doesn't protect the profession; it isolates it.

To move forward, EMS must invest in digital readiness, demand vendor accountability, and develop leaders fluent in innovation, not just operations. This is not about tech—it is about shaping the future before it is shaped for us.

3

Lessons from Visionaries in Emergency Medical Services

LEARNING OBJECTIVES

After reading this chapter, the reader will be able to

- Analyze the historical development of EMS in the United States by identifying the key contributions of visionary leaders who helped shape national EMS systems and standards.
- Evaluate the leadership principles demonstrated by early EMS founders—curiosity, collaboration, excellence, standards, adaptability, inclusivity, and innovation—and apply them to current challenges in EMS practice and system design.
- Interpret the relevance of historical EMS leadership strategies to contemporary issues, including the integration of AI, system interoperability, and national consistency.
- Differentiate between ad hoc innovation and coordinated visionary leadership, using historical case examples to demonstrate how structure, principles, and consensus accelerated EMS professionalization.
- Formulate strategies for modern EMS leadership that incorporate legacy leadership principles into future-facing initiatives, particularly those involving emerging technologies such as AI.

A Vision Worth Building

In 2005, I received a phone call that would alter the course of my life. The voice on the other end described the devastating toll of the 2004 Indian Ocean tsunami in Sri Lanka—a disaster that had claimed tens of thousands of lives in the island nation. The call included a bold invitation:

> *Would I come for six months to help improve the EMS system in Sri Lanka?*

At the time, I imagined I would be consulting on protocols, refining training standards, or evaluating existing ambulance operations. But when I landed in Sri Lanka, the reality hit hard—there was no national EMS system to improve. It had never existed in the first place. There were no paramedics, emergency medical technicians (EMTs), or national emergency numbers. The concept of an ambulance was a stripped-down box van with a stretcher and a red light—no medical supplies, no trained personnel, and no expectation of clinical care during transport.

I had come to help upgrade a system, but instead, I faced the challenge of building one from scratch.

What was intended to be a six-month leave of absence became nearly a decade of relentless work. It would have been easy to walk away. The task was overwhelming, the infrastructure did not exist, the workforce did not exist, and the cultural expectations for EMS had not yet been formed, but the vision—the *need*—was undeniable.

We did not build the system overnight. We built it together—slowly, deliberately, and in partnership with local leaders. Over time, Sri Lanka developed a national EMS system serving a population of 22 million across multiple languages. Thousands were trained as EMTs. Modern ambulances were fielded with equipment that sometimes exceeded the technology used in American ambulances. We also built a single, technologically advanced, unified national dispatch center that coordinated care across the island nation.

The work was challenging and required compromise. Unscalable ideas had to be set aside. Budgets and priorities shifted, but one thing never changed: We never compromised on standards.

We demanded excellence, remained adaptable, inclusive, invited voices who did not always agree, continuously innovated, and remained clear on our shared mission.

Years later, the World Bank would cite Sri Lanka's EMS system as one of the most advanced in the developing world.

You might be wondering why I've started this chapter here. In EMS, we often think vision is reserved for the privileged—those with funding, time, or power. I have heard it from every corner of our profession:

> *That is nice, Donnie. But we're selling cupcakes to buy defibrillators.*
>
> *We cannot afford that kind of future.*

Let me be clear: Vision does not require money. It requires courage.

When Dr. Peter Safar wrote about a nationwide computerized EMS reporting system in 1969, the only computers were at large universities, big private corporations, or in a few government systems, like NASA. His idea was not remotely practical then, but he wrote it anyway. Visionaries do not wait for the world to be ready—they get the world ready.

This chapter is about those visionaries—the men and women who did not just dream of a better EMS system; they built one, not for themselves, not for their city, but for a nation. And now, it's our turn.

Built Together: The Founders of Modern Emergency Medical Services

The American EMS system did not emerge from a federal mandate or corporate investment. It was built patiently and deliberately by individuals who believed in something larger than themselves.

From the late 1960s through the 1980s, a small group of clinicians, educators, public servants, and medical pioneers transformed a patchwork of funeral home transports and volunteer rescue squads into a recognized profession with national standards.

These visionaries had every reason to act alone. Many were already respected experts—physicians, researchers, and policymakers with the stature to drive local reform on their own terms. But they chose a more difficult path: They worked together, debated, and compromised. They laid the foundation for a national EMS system designed to serve every community in the country.

That collaboration was not easy. Some ideas were set aside—not because they lacked merit, but because they could not be done to scale. Some proposals moved slowly, not for lack of urgency, but because consensus took precedence. And yet, through that friction, something remarkable was forged: a profession united by standards, grounded in science, and committed to public service.

The result was not just a more organized ambulance system. It marked the creation of a new branch of medicine—prehospital emergency care—built from scratch by people who refused to settle for chaos.

Now, as we confront the rise of AI, we would remember what guided these early leaders: a commitment to curiosity, collaboration, excellence, and adaptability. These were not abstract ideals. They were practical strategies—operational blueprints that transformed chaos into a profession. Today, they remain just as essential.

These are not nostalgic virtues. They are operational imperatives.

As AI begins to touch every aspect of EMS—from documentation to diagnostics—these same principles must guide us again. What follows is not a list of names; it is a blueprint.

Visionary Leaders of Emergency Medical Services: Selected Pioneers Who Helped Build a National Profession

The rise of modern EMS in the United States did not result from any one person, program, or policy. It was the outcome of many visionary leaders—clinicians, educators, policymakers, and advocates—who saw the potential for a coordinated, lifesaving system and worked across barriers to bring it into being.

The following individuals represent just a few of those transformative leaders. Their contributions illustrate the diversity of talent and commitment that shaped the profession. While countless others also played critical roles—especially at the local, regional, and state levels—these examples capture the spirit, strategy, and courage that continue to guide EMS into the future.

Dr. Peter Safar, the "father" of CPR and critical care medicine, shaped the modern EMS concept. He championed the integration of field care, led the development of paramedic education through the Freedom House Ambulance Service, and envisioned a data-driven EMS system years before the technology existed.

Dr. Charles Rockwood was the driving force behind organizing the EMS program in San Antonio, Texas, in the early 1970s, serving as its first medical director. This initiative transformed emergency care in the city. Moreover, the textbook he helped publish in collaboration with Dr. Farrington and others, *Emergency Care and Transportation of the Sick and Injured*, became the first standard reference for EMT training curricula nationwide. Published by the American Academy of Orthopedic Surgeons, this textbook, known as the "orange book," remains essential today.

Dr. JD "Deke" Farrington pioneered prehospital emergency care, developing the first formal training programs for EMTs. His work emphasized the importance of structured education and continuous professional development. His course outline was later transformed into his book, *Emergency Care and Transportation of the Sick and Injured.* Dr. Farrington served as the first physician Chair of the National Registry of EMTs board of directors. In 1975, he called upon the medical establishment to officially recognize EMT-paramedics as a bona fide allied health medical profession. His contributions laid the groundwork for the professionalization of EMS and set high standards for training and practice.

Dr. Norman McSwain led the establishment of a pilot paramedic program in Kansas before the national standard. He later revolutionized trauma care by developing prehospital trauma life support courses, significantly improving how EMS clinicians approach trauma and enhancing patient survival rates. His work in trauma care has had a lasting impact on EMS practices and training programs worldwide.

Dr. Eugene Nagel integrated advanced medical techniques into prehospital care, advocating for paramedics' use of defibrillators and advanced airway management. His efforts paved the way for EMS to provide more advanced care in the prehospital setting.

Dr. Nancy Caroline was a trailblazer in EMS physician-led medical direction and EMS education. Her textbooks and training programs set the national standard for paramedics, ensuring EMS clinicians had the knowledge and skills to save lives. Dr. Caroline also played a significant role in forming the Freedom House Ambulance Service, working alongside Dr. Safar.

James O. Page, JD, was an EMS management and advocacy visionary. He founded the *Journal of Emergency Medical Services* and was a relentless advocate for EMS professionalization, pushing for legislative and organizational advancements. His work has profoundly impacted the development and recognition of EMS as a vital component of health care.

Dr. David Boyd, a pioneer in trauma and EMS, conceptualized and established trauma units that combined monitoring, resuscitation, and immediate surgery. His work led to the development of specialized statewide trauma centers and the implementation of a computerized trauma registry, significantly advancing trauma care.

Roddy A. Brandes, an often under-recognized figure, established Mecklenburg Emergency Services in Charlotte, NC, and played a pivotal role in advocating for the standardization and enhancement of EMS. His testimony before Congress and subsequent actions led to the formation of the National Highway Traffic Safety Administration and national EMS guidelines and standards. Roddy was a prominent leader of the Ambulance Association of America and the board's first chair for the National Registry of EMTs.

Rocco Morando, an EMT training coordinator at Ohio State University, realized the urgent need for national standards and a way to assess the entry-level competency of EMTs and paramedics against a national standard. Morando collaborated with the visionaries and became the first executive director of the National Registry of Emergency Medical Technicians, ensuring that EMS clinicians met a consistent, professional, and independently validated standard nationwide.

These pioneers embraced essential principles today: curiosity, collaboration, excellence, shared standards, adaptability, inclusivity, and continuous innovation. Their stories reflect more than personal achievement—they represent a collective movement that turned scattered rescue services into a national profession.

As EMS confronts a new era of transformation, driven by AI and emerging technologies, these principles offer a roadmap. The leaders profiled here are not the only ones who helped build EMS; their legacies remind us that vision and action can change the future. Let their example guide the next generation of EMS clinicians and leaders forward.

Core Principles of Visionary Leaders

How the Founders of EMS Built a System and What We Must Learn from Them

What set these pioneers apart was not their titles—it was how they led. Across backgrounds and regions, they demonstrated shared principles that transformed a fragmented patchwork of rescue efforts into a unified national profession.

These were not abstract ideals. They were strategic behaviors—tested under pressure and refined through collaboration. As EMS faces another turning point, with AI reshaping documentation, diagnostics, and operations, those same principles must guide us again.

If there is one enduring lesson from the early builders of EMS, it is this: Progress does not happen by accident. It happens through principle-driven leadership.

What follows are seven core principles—practical, field-tested tools that powered the creation of modern EMS. They remain just as urgent today as we navigate the subsequent transformation of our profession.

1. **Embrace Curiosity**

   Progress starts with a question. The early EMS leaders were not satisfied with the way things were. They asked, *What if we could do better? What if the ambulance was more than a ride? What if care could begin before the hospital doors?*

   That curiosity led to new protocols, technology, and a new profession.

   Today, we must apply the same mindset to AI. We need to start asking how these tools might reduce diagnostic errors, how predictive algorithms might improve outcomes, and how intelligent documentation might minimize burnout. Do not wait for perfect answers—start by asking better questions.

2. **Work in Collaboration**

   The EMS founders worked across agencies, regions, and disciplines. They did not wait for perfect alignment—they *built* it.

   Many could have carved out elite local programs and called it a day. Instead, they chose the more challenging path of national collaboration.

   If EMS is going to integrate AI responsibly, it will not happen in silos. It will take collaboration between clinicians, technologists, regulators, educators, and—importantly—each other. EMS must speak with a unified professional voice to shape our future.

3. **Demand Excellence**

   There was nothing casual about how EMS was built. The founders wrote rigorous training standards, evaluated outcomes, and built national certification systems—not because it was easy, but because excellence protects patients.

   AI demands the same standard. These systems must be vetted, validated, and continually refined. Half-measures in AI implementation are not only unprofessional—they are dangerous. Leaders must reject shortcuts and insist that innovation enhances quality, not replaces it.

4. **Value Standards**

   Uniform standards were the glue that held the early EMS system together. The first national curricula, treatment protocols, and licensure models did not just raise the floor—they defined the profession.

   With AI, standardization is again nonnegotiable. We must ensure interoperability between systems, consistency in deploying tools, and equity in accessing them. A patchwork of local experiments without national cohesion will set us back.

5. **Adapt to Change**

   The EMS pioneers did not cling to outdated ideas. They changed protocols, embraced new technologies, and restructured organizations. They understood that adaptability was not weakness—it was leadership in action.

   AI will require the same flexibility. As capabilities evolve, our systems must evolve with them. Protocols, scope of practice, and clinical training are all on the table. Resistance to change will not protect us; strategic adaptation will.

6. **Foster Inclusivity**

   From Freedom House to rural pilot programs, the history of EMS includes leaders who actively brought in previously excluded voices. The result was not just political—it was operational. Inclusivity made EMS better, stronger, and more trusted.

   AI systems must be built the same way. Inclusion in design, testing, deployment, and governance is ethical and practical. Tools trained on narrow datasets and deployed without input from diverse communities will fail the people we are sworn to serve.

7. **Innovate Continuously**

   The original EMS visionaries never stopped pushing the envelope. From defibrillators in the field to paramedic-level airway management, their mindset was always about moving forward.

   That spirit must continue with AI. Leaders must cultivate an innovation mindset—not just adopting tools but continually asking how they can be refined, reimagined, or replaced. Continuous improvement is not an add-on—it is the job.

These principles are not outdated ideals. They are your operational playbook.

The same mindset that launched modern EMS will guide us through this new era of complexity, data, and intelligent systems. The tools are different, but the stakes are the same. And the mission—serving patients with speed, safety, and skill—has never been more urgent.

The Seven Principles That Built Emergency Medical Services and Will Shape its Artificial Intelligence Future

TABLE 3–1. The principles that built modern EMS with the leadership actions needed today as AI transforms the field.

Principle	Then: How EMS Was Built	Now: What EMS Needs for AI
Curiosity	"What if the ambulance could be the beginning of care?"	Ask how AI can reduce errors, anticipate needs, and improve accuracy.
Collaboration	National leaders compromised to build shared standards.	EMS must unify—across agencies and disciplines—to shape AI policy.
Excellence	Training, protocols, and certification were not optional.	Demand rigorous testing and validation of every AI tool used.
Standards	National curricula and licensure created cohesion.	Require interoperability and ethical use across platforms.
Adaptability	Pioneers changed course when needed to stay effective.	EMS systems must evolve with AI—quickly and strategically.
Inclusivity	Voices from underserved communities helped build trust.	Ensure AI reflects and serves all populations equitably.
Innovation	EMS leaders pushed for the impossible—and achieved it.	Keep refining AI tools. Stay curious. Lead the next transformation.

These are not just historical values—they are operational imperatives. Use them. Teach them. Lead with them.

Implementation and Execution of the Vision

A vision without execution is just a good idea.
Good ideas do not change the world.

The founders of EMS did not merely imagine a better future—they built it. And what made the difference was not just intellect or inspiration. It was action: deliberate, coordinated, uncompromising action.

They did not wait for perfect conditions. The conditions were terrible; there was no national curriculum, no federal support, and no cultural consensus that EMS was even a profession. Despite these obstacles, they moved forward, and they did not do it alone. They aligned around shared principles, worked across silos, and executed relentlessly.

One of the clearest examples comes from Dr. Peter Safar. In 1969, he submitted a proposal to the Airlie House Conference calling for a standardized, computerized EMS patient care reporting system. It was decades ahead of its time. Back then, computers cost millions and were only accessible to institutions like NASA. Safar knew this. He also knew that visionary infrastructure requires foresight. He was not solving for *today*—he was laying the foundation for what EMS could *become*.

That proposal became the philosophical spark for what we now know as NEMSIS—the National EMS Information System. While Safar did not live to see the whole system deployed, his vision was clear: Data saves lives when structured, shared, and used to improve care.

Other leaders followed the same model:

- Dr. Charles Rockwood helped launch one of the first citywide EMS systems and coauthored the first national EMT textbook.
- Dr. Farrington developed standardized training and pushed for the recognition of EMS as an allied health profession.
- Rocco Morando established the National Registry of Emergency Medical Technicians (National Registry) to ensure clinical competency across state lines.

None of this was easy. Standards were debated. Progress was slow. And every advancement came with resistance, from within the system and from those who did not yet see EMS as essential. However, what made these leaders successful was not that their ideas were perfect. It's that they executed with urgency and resolve.

They understood something modern EMS leaders must not forget: You do not wait until you have buy-in from everyone. You lead with clarity, bring others along, and build toward the future—one standard, one program, one patient at a time.

And now, it's our turn.

The stakes are just as high as AI reshapes documentation, diagnostics, training, and system design. EMS leaders today must implement and execute just as fiercely as those who came before. That means

- designing AI pilots that are clinically sound and ethically grounded,
- vetting vendors, not just for their promises but for their proven outcomes,
- training clinicians on how to use (not fear) AI-enabled tools, and
- decisively—not recklessly—moving forward.

The past did not build itself. The future will not build itself either.

Applying the Core Principles of Visionary Leadership Today

EMS leaders today stand at the edge of a new frontier. AI is not a distant concept—it is already in the clinical tools, billing systems, dispatch platforms, and electronic patient care reporting we use daily. What remains uncertain is whether EMS will shape the role of AI, or allow AI to shape its role.

To lead well in this moment, we must return to the same core principles that built EMS from the ground up. These principles are not artifacts—they are active tools. They remain just as relevant today as they were in the 1960s. And they demand application, not admiration.

Embrace Curiosity

Curiosity drove the birth of modern EMS, and it must now drive the integration of AI. You must ask questions others are not asking yet:

- How can machine learning reduce scene times without compromising care?
- How do we train paramedics to use diagnostic decision support safely?
- Where are the blind spots in today's AI tools?

Leaders who ignore AI will find themselves reacting to its consequences. Leaders who engage with curiosity will shape their future.

Work in Collaboration

This is the moment for EMS to unify—not just within individual services or states, but as a profession. We must develop a national voice on AI integration. This means convening clinicians, educators, regulators, developers, and patients to design solutions that work in the field, not just in theory.

Collaboration is just as critical inside your organization. Your IT team, clinical educators, and frontline crews must not be siloed. If they cannot talk, your AI strategy will fail before it starts.

Cross-functional collaboration is no longer optional—it is operationally essential.

Demand Excellence

The AI tools flooding the market promise faster documentation, better triage, and more innovative system design; however, promises are not enough. You must demand proof.

Before you adopt an AI product, there are questions to consider:

- Has this tool been validated in the EMS environment?
- How was the training data selected? Who was excluded?
- What's the failure mode—and who is accountable?

Excellence is not about being perfect. It's about refusing to accept mediocrity, especially when lives are on the line.

Value Standards

AI will expose the cracks in EMS standardization if we're not careful. One system cannot use a 12-lead electrocardiogram interpretation algorithm, while another uses a different version with no audit trail, and both report to the same quality assurance board. We have seen what fragmented systems do to EMS credibility.

National standards and common frameworks must govern AI. Without them, we will return to the pre-1970s era, where outcomes depended on zip codes, not protocols.

Set standards. Enforce them. Expect others to do the same.

Adapt to Change

You do not need to have all the answers, but you do need to move with the times.

Leaders who resist change will be left behind. The public will not wait. Payers will not wait. Hospitals will not wait. And your workforce—especially the next generation—expects you to lead into the future, not defend the past.

Adaptability is not surrendering. It's a strategy. The best leaders flex without breaking.

Foster Inclusivity

If EMS is to serve the public, then the tools we deploy—especially those powered by AI—must be shaped by diverse voices:

- Include underrepresented populations in AI datasets.
- Listen to the clinicians who know what the tools do in the back of a rig.
- Engage the community before automation creates inequity.

If AI makes EMS faster but not fairer, we have failed.

Innovate Continuously

Innovation in EMS has never been a one-time leap. It's been a steady climb. As AI enters our space, we cannot adopt it once and stop. Leaders must

- pilot tools,
- refine workflows,
- collect feedback,
- update policies, and
- repeat continuously.

Innovation is not a department. It's a discipline.

The bottom line is that EMS cannot afford passive leadership in the face of technological transformation. The same principles that built a national profession out of chaos must now guide the integration of AI into everything we do—from training and documentation to diagnostics and response models (fig. 3–1).

The next decade will define whether EMS becomes more intelligent, agile, and respected or whether we miss the moment entirely.

The leaders who built EMS did not wait for the future to arrive—they built it. Now it's our turn. If we do not lead the integration of AI, someone else will.

FIGURE 3–1. Visionary leadership in EMS demands both courage and discipline. By fostering curiosity, collaboration, inclusivity, and innovation, leaders can guide their teams through rapid technological change while maintaining excellence, shared standards, and public trust.

Key Terms and Definitions

- **EMS Visionaries:** Pioneering leaders from the 1960s to 1980s who were foundational in building a coordinated, standards-based EMS system in the United States through education, policy, clinical practice, and system design.
- **Freedom House Ambulance Service:** The first program in the United States to train paramedics at a high level of medical care, launched in Pittsburgh, PA, in the late 1960s. It served as the model for the national paramedic curriculum.
- **Airlie House Conference:** A 1969 gathering of EMS experts and policymakers convened to define the structure of national emergency medical services. It catalyzed future EMS development.
- **National Standard Curriculum:** A federally supported curriculum created in the 1970s to standardize EMT and paramedic education nationwide, ensuring consistent training and clinical expectations.
- **National Registry of EMTs** is an independent, nonprofit organization certifying EMS clinicians through standardized examinations. It remains the principal national certifying body in EMS.
- **Prehospital Trauma Life Support:** A standardized training program developed by Dr. Norman McSwain to improve trauma care delivered by EMS clinicians, now used globally.
- **EMS System Design:** The structural, operational, and clinical configuration of how emergency medical care is delivered, including dispatch, clinical protocols, training, quality oversight, and integration with public safety and hospitals.
- **Visionary Leadership:** Leadership defined by principle-driven decision-making, long-term thinking, and the courage to challenge the status quo to advance a profession, especially under conditions of uncertainty or adversity.

Reflection Questions

1. What barriers in my agency or system am I accepting today that visionary leaders before me would have challenged—or changed?
2. Am I building for tomorrow or simply managing today? What percentage of my time is spent preparing my team for the future versus reacting to short-term pressures?

3. Who am I collaborating with outside my department, region, or discipline? If the answer is "no one," what am I doing to change that?
4. How am I holding myself—and the technologies I use—to the standard of excellence and patient protection our founders demanded?
5. If someone looked back on my leadership 20 years from now, what would they say I helped build? Would they see a caretaker of tradition—or a builder of the future?

Study Questions

1. What leadership mindset did the author adopt when asked to help build an EMS system in Sri Lanka—a country with no infrastructure, workforce, or prehospital care tradition?
2. What was the most significant challenge faced upon arriving in Sri Lanka to assist with EMS?
3. Visionary leadership in EMS does not require ____________; it requires courage.
4. Which of the following is NOT one of the core leadership principles discussed in the chapter?

    a. Collaboration
    b. Regulation
    c. Inclusivity
    d. Adaptability

5. What was Dr. Peter Safar's vision submitted at the 1969 Airlie House Conference?
6. Name two contributions of JD "Deke" Farrington to the development of EMS.
7. How did the Freedom House Ambulance Service impact EMS education and equity?
8. Why are national standards and interoperability critical in the era of AI, according to the chapter?
9. Describe how the leadership actions of EMS founders like Rockwood and Morando mirror the needs of today's leaders facing the growing inclusion of AI.
10. What is the primary message this chapter conveys to current EMS leaders?

CHAPTER SUMMARY

Learning from the Past, Leading into the Future

This chapter reflects on the leaders who transformed American EMS from fragmented services into a coordinated national system—and distills the principles that enabled their success. Through stories of visionaries like Safar, Farrington, Rockwood, and Caroline, we uncover a blueprint for EMS innovation grounded in curiosity, collaboration, standards, and execution. These same principles must now guide the profession as it confronts a new wave of transformation driven by AI. Leadership today must be as bold, intentional, and unrelenting as the pioneers who came before us.

The EMS system we know today was not handed down—it was built. Not by chance, not by convenience, but by visionary leaders who worked together, compromised when needed, and pushed forward with discipline and purpose.

Before their efforts, there was no EMS profession: no national curriculum, no certification standards, no universal training or licensing, and no consistent emergency response infrastructure; yet, by applying seven core principles—curiosity, collaboration, excellence, standards, adaptability, inclusivity, and continuous innovation—they created a profession that now saves hundreds of thousands of lives every year.

They acted with a bold vision and national focus. They did not settle for what was easy, cheap, or politically convenient. Perhaps most impressively, they chose to build something bigger than themselves.

This chapter is not just a history lesson. It's a challenge. As AI reshapes how we assess, document, diagnose, and deploy care, you are in the same position those early pioneers once held. The decisions you make today—about what to adopt, what to resist, and how to lead—will determine whether EMS grows stronger or becomes irrelevant.

We do not need a million-dollar budget to lead.

We do not need a tech degree to ask good questions.

We do not need permission to start building something better.

EMS leaders must have courage and clarity. Most importantly, we must take action. The founders of EMS did not wait for someone else to lead, and neither should we. Our generation of EMS leaders must rise to this moment with clarity, courage, and purpose.

4

Barriers to Innovation

Challenges Facing Emergency Medical Services

LEARNING OBJECTIVES

After reading this chapter, the reader will be able to

- Identify the systemic, cultural, and regulatory barriers historically hindering innovation in EMS, including underfunding, fragmentation, and risk-averse leadership structures.
- Analyze the role of cognitive biases, such as anchoring, groupthink, and the sunk cost fallacy, in shaping resistance to emerging technologies and innovation within EMS agencies.
- Evaluate how outdated policies and rigid regulatory frameworks can limit the adoption of AI and other emerging technologies, and articulate strategies for adaptive regulation.
- Apply behavioral science and change management principles to design practical approaches that overcome resistance to innovation and foster a culture of evidence-based leadership.
- Compare EMS's current response to emerging technologies with historical examples of industries that failed to adapt to disruptive change, such as the BlackBerry/iPhone case study.
- Formulate a leadership strategy for responsibly guiding EMS innovation, including piloting new tools, managing staff resistance, ensuring equity, and building systems resilient to bias. Critically reflect on their organizational posture toward innovation, and assess whether their system is positioned to lead, lag, or be left behind in the AI-driven healthcare transformation.
- Evaluate AI deployment's legal and ethical implications in EMS, including privacy, liability, informed consent, and data security concerns.
- Explain the value of model rules and national governance frameworks in reducing legal ambiguity, ensuring ethical AI integration, and preventing state-by-state fragmentation.

Prologue: Paper Cuts and Pushback

The year was 2014, and I had just been appointed state EMS director for Louisiana.

During my final interview, the assistant secretary of health described the EMS office as "a little outdated regarding technology." That turned out to be an understatement.

When I walked into the office, I was greeted by a stack of paper license applications, measured not in inches but in feet. Hundreds of envelopes arrived daily from the post office. Inside? Handwritten forms, money orders, personal checks, and sticky notes. It was a scene frozen in time.

With a smirk, one staff member gave me the "good news": I could set up a cot in a free office. Why? Because every year, during the last two weeks of March, the staff worked around the clock to process paper renewals—sleeping at the office, hand-sorting paperwork, and stuffing envelopes. I was informed that I'd be expected to hand-sign every emergency medical technician (EMT) and paramedic license personally (fig. 4–1).

There was no time to fix the system on the spot. We got through the cycle—every application received before the deadline was processed, and every card was mailed. But as soon as that wave passed, I made one commitment: Never again!

We would implement a digital licensing system. No more paper. No more cash. No more mailing licenses by hand. That is when the real resistance began.

Some staff members supported the move; others quietly hoped I would give up:

- "People will not trust an online system."
- "This is how it has always been done."
- "They will want that printed card for their wallet."

One EMS director told me flatly, "Donnie, we are a small service. Most of my volunteers do not even own a computer, let alone have email. Your system will never work for us."

I promised, "If you have a clinician who cannot access the system, give them my direct cell number. I'll help them myself."

When we launched the new system, I got one call. It was an 80-year-old EMT who needed to renew his license. Since he did not have an email address, I gladly helped him.

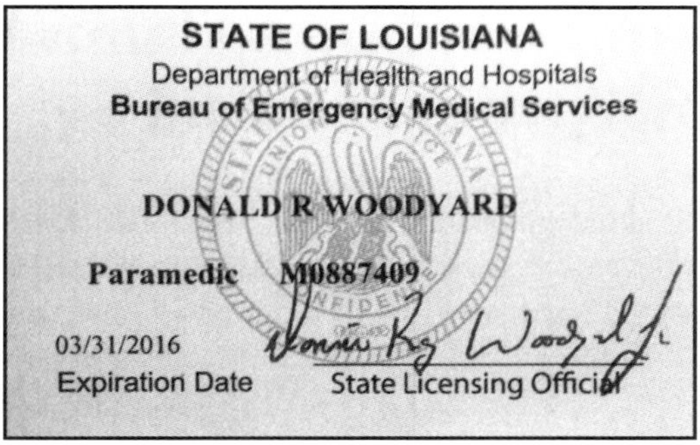

FIGURE 4–1. Donnie's 2014 hand-signed paramedic license, a reminder of a time not so long ago. Today, licenses are issued electronically. This is a small but telling example of how even professional identity has become digital.

That was the only call.

Turnaround times dropped from weeks to hours. Error rates plummeted. License verification became transparent and publicly accessible for the first time. There is not one complaint about the lack of a hand-signed paper wallet card.

Everyone says they want change. But when change shows up, almost everyone resists it.

That is the paradox. Every EMS leader must learn to navigate it, especially now, as AI forces us to rethink everything from training to triage.

But the resistance is not just cultural. The barriers to innovation in EMS are also structural, ethical, legal, and operational. Fragmented governance slows progress. Risk-averse leadership misses opportunities. And outdated regulatory frameworks—written before the arrival of dynamic, self-learning algorithms—are now being asked to govern tools they were never designed to handle.

We face a widening gap: AI vendors are moving fast, and patients are expecting more, but EMS is constrained by systems that lag behind, not just technologically but conceptually.

Being compassionate does not mean being passive. Listening matters, adjusting matters, but so does leading. In EMS, leading change means understanding the barriers you are about to encounter—from liability fears to policy inertia to clinician distrust—before they stop you cold.

Systemic Barriers to Innovation in Emergency Medical Services

When people talk about innovation in EMS, they often focus on tools—new monitors, airway devices, computer-aided design upgrades, or digital charting platforms; however, the real barriers to progress are not technical. They are structural.

1. Underfunding and Misaligned Incentives

EMS remains chronically underfunded—not just at the agency level but across the entire infrastructure. We are funded to transport, not to treat. We are paid to show up, not to prevent. We teach innovation in theory, but reward the status quo in billing and regulation:

- What incentive does a service have to invest in predictive analytics when reimbursement is tied to mileage, not outcomes?
- We are asked to do more with less and punished when innovation disrupts a fragile funding model. Leaders cannot afford to ignore this tension, but cannot use it as an excuse to stand still.

Even when EMS leaders recognize the value of innovation, outdated infrastructure often stands in the way. Many agencies still operate on legacy ePCR systems, disconnected computer-aided design platforms, and fragmented dispatch protocols. These systems were not designed to support AI integration; in many cases, they actively inhibit it. EMS systems cannot fully leverage AI's potential without modern application programming interfaces, real-time data flows, or cloud-based flexibility. Closing this gap requires more than new software—it demands strategic investment in digital infrastructure, vendor accountability, and state-level support for interoperability and modernization.

2. Fragmentation Across States and Systems

The repeal of the federal EMS Act and the end of dedicated federal funding in the early 1980s led to persistent fragmentation within EMS. This fragmentation, often summarized by the saying, "If you've seen one EMS system, you've seen one EMS system," significantly hinders innovation.

This fragmentation paralyzes innovation. A brilliant AI triage tool in one state could be illegal in its neighboring state, and a regional dispatch enhancement in Michigan may be incompatible with systems in Missouri.

Innovation does not scale when every border is a barrier. This is why initiatives like the United States EMS Compact and National EMS Information System are critical—they are not just data-sharing or mobility platforms. They are the scaffolding for national innovation.

3. Leadership Gaps and Risk Aversion

Let's be blunt: Many EMS agencies are still led by well-intentioned field clinicians who were promoted for being good medics, not because they were trained to navigate innovation, budgeting, or strategic planning.

Without an exemplary leadership mindset, innovation feels like a risk, not an opportunity:

- What if it fails?
- What if the staff pushes back?
- What if the regulators do not approve?

Hesitation is understandable—but it is not our legacy. EMS has never waited for perfect conditions. Even in uncertainty, forward motion is how we reclaim leadership and shape outcomes.

4. Outdated Infrastructure and Training Limitations

Even when leaders are willing to innovate, the foundation they work with may not support the weight of modern tools. Many EMS agencies operate with outdated dispatch software, legacy ePCR systems, and underpowered mobile devices that cannot reliably run AI-enabled applications. Even basic connectivity or device compatibility may hinder rural and resource-constrained systems.

Technical limitations are only part of the problem. Equally urgent is the workforce readiness gap.

Most EMS clinicians and agency leaders have never received formal education on AI, digital systems integration, or basic data literacy. EMS educators often lack access to contemporary teaching tools, and most national training curricula still treat technology as an add-on, not a core competency. This creates a dangerous loop: Systems hesitate to adopt AI because the workforce is unprepared, and the workforce remains unprepared because the system will not adopt.

If EMS is going to engage with real-time clinical decision support, predictive analytics, or AI-enabled documentation, it must first invest in foundational training and infrastructure upgrades. That includes

- equipping field clinicians with devices that can support advanced tools,
- upgrading backend systems and connectivity,

- ensuring secure HIPAA-compliant data pathways, and
- integrating AI and digital literacy into initial education and continuing education programs.

EMS agencies risk falling further behind without these investments—not because they lack vision, but because their systems were never designed to evolve (table 4–1).

Innovation is not blocked by tools—systems block it, and systems do not change until leaders do.

TABLE 4–1. Systemic Barriers That Keep EMS Stuck

Barrier	Why It Matters
Underfunded & Misaligned	EMS is paid to transport, not transform. Innovation often disrupts the current revenue model.
Fragmented & Inconsistent	Fifty states. Hundreds of protocols. No unified path forward. Innovation cannot scale in silos.
Risk-Averse Leadership	Too many systems fear failure more than stagnation. Playing it safe means falling behind.
Outdated Infrastructure & Training Gaps	Many EMS agencies lack the technology and workforce readiness to support modern tools. Innovation fails without reliable systems and educated users.

Cultural Resistance and Fear of Change

Every EMS leader has heard it:

- "We have always done it this way."
- "My people will never go for that."
- "That is fine for big-city systems, but it will not work here."

Let's call it what it is: fear masquerading as realism.

In EMS, we pride ourselves on adaptability in the field. But inside our organizations, change is often met with skepticism, sarcasm, or outright hostility. And that resistance does not just come from the front line. It comes from seasoned leaders, credentialed educators, and long-tenured clinicians—those who should know better.

The Waymo Moment

In late 2024, I posted a short piece on LinkedIn about autonomous vehicles and their inevitable role in EMS. I highlighted how companies like Waymo were already operating driverless taxis in significant cities and suggested that it was only a matter of time before this technology expanded into last-mile delivery, over-the-road trucking, and, eventually, ambulance operations.

The response was swift and visceral.

One EMS leader replied with a meme: a photo of a Waymo vehicle confused by a traffic cone (fig. 4–2). Others piled on. Some questioned the technology. Most questioned my sanity. To them, the idea of AI-driven ambulances was not just far-fetched—it was threatening.

Shortly afterward, I did a video interview with the National Registry of EMTs on the same topic. It went viral—over 100,000 views in just a few days. But the majority of comments were not celebratory. They were defensive, angry, and dismissive. That is the lesson: In EMS, resistance does not always come from evidence. It comes from *identity.*

To many EMS professionals, the ambulance is not just a vehicle. It's a symbol of control. Suggesting that software could drive—or even assist—feels like an attack on our relevance. So, instead of engaging critically, people mock, reject, and retreat.

FIGURE 4–2. Autonomous vehicle technology, once experimental, is now operating on public roads. The same principles enabling driverless cars will soon be incorporated into EMS logistics, fleet safety, and even autonomous patient transport.

But here's the truth:

- Waymo vehicles have already driven millions of miles in major U.S. cities with a fraction of the crash rate of human drivers.
- Autonomous transport is already intersecting with healthcare logistics.
- This is not theoretical—it's happening now.

You do not have to love it. But you do have to lead through it. EMS professionals have the power to shape the intersection of these technologies with our field, ensuring that innovation serves our unique needs and enhances our capabilities.

Ethics, Liability, and the Case for Model Rules

Regulation protects the public, but regulation alone is not enough. EMS leaders must also confront the ethical and legal responsibilities of AI integration. Unlike traditional protocols or tools, AI systems can continuously learn, evolve, and make real-time recommendations. That raises urgent questions:

- Who is accountable when a machine-generated recommendation results in harm?
- How do we ensure that patient data collected through AI is kept secure and private?
- What does informed consent mean when clinicians may not fully understand how the algorithm works?

These are not abstract concerns. They are real legal risks. If EMS agencies deploy AI tools without apparent oversight, robust business associate agreements, and protections for patient privacy under HIPAA and state laws, they could face lawsuits, reputational damage, and patient harm.

Beyond liability, ethical vigilance matters. EMS leaders must demand transparency from AI vendors. They must ask: Who trained the model? What biases are embedded in the data? Has this tool been validated in diverse communities—urban and rural, high-resource and low-resource, across all patient populations?

Equity is not optional. If EMS adopts tools trained mostly on urban data or adult populations, the result could be under-triage in rural counties or misdiagnosis in pediatric and geriatric patients. Ethical AI requires representation, transparency, and explainability.

To manage these risks at scale, EMS must follow the path carved by other professions: coordinated governance.

Model legislation and model rules are essential. The EMS Compact (Recognition of EMS Personnel Licensure Interstate CompAct [REPLICA]) offers a proven example. By aligning states through standardized law, REPLICA has created a seamless, legally sound mechanism for interstate clinician practice. That same structure can be adapted for AI, ensuring that AI tools deployed in EMS meet minimum ethical, legal, and performance standards, regardless of jurisdiction.

The time to act is now. Waiting for 50 separate states to regulate EMS AI tools independently will create confusion, delay, and risk. National alignment is not just efficient but a moral and operational imperative.

Policy and Regulation as Double-Edged Swords

In EMS and across health care, regulation is both a shield and a challenge. It protects the public, establishes baseline competency, reinforces accountability, and ensures a consistent standard of care, especially in a profession that operates in private homes, behind closed ambulance doors, and in life-or-death moments with limited oversight.

But that same system can also inhibit progress—not out of neglect, but because many structures were built for a different era. Regulations that were essential in the past often struggle to keep pace with present-day complexity and future-facing tools.

For example, a 2005 policy requiring medical directors to approve every care algorithm made perfect sense in a world of static protocols. But what happens when the "algorithm" is an evolving machine learning model responding to live patient data in real time? Is it reviewed line by line? Reapproved annually? Validated manually?

Most regulatory frameworks were designed to detect fraud, prevent harm, and ensure accountability. That mission is essential. But as technology changes how care is delivered, regulation must expand—not just to govern innovation, but to enable it.

The Innovation Gap

The result is a growing innovation gap:

- Vendors move fast.
- Clinicians adapt quickly.
- Patients expect more.
- Regulators struggle to keep up.

This gap carries real consequences—not only for innovation, but for clarity and confidence. EMS leaders may hesitate to adopt promising tools, vendors may delay partnerships, and ultimately, patients may lose access to advancements that could save time, resources, or lives.

Regulation should never be eliminated, but it must be modernized.

It is time to move forward:

- Build adaptive policy models that support real-time learning systems
- Create innovation pathways for EMS, modeled after the Food and Drug Administration's approach to digital health tools
- Involve regulators early in design conversations, not just approval processes
- Train emerging EMS leaders in regulatory innovation, not just compliance

To lead effectively, EMS must also speak with a unified national voice. Model legislation, national administrative rules, and coordinated policy frameworks are not optional but essential infrastructure for responsible AI integration.

Fortunately, we have a precedent. REPLICA showed how shared governance across states can improve mobility, safety, and accountability. A similar strategy can guide AI's ethical and practical use in EMS, ensuring that 50 different regulatory interpretations do not obstruct clinical progress.

Moving Forward: Designing the Path to Lead

EMS risks being held at the intersection without modernized, agile regulation, waiting while other sectors advance through the green lights of innovation and investment. This is not because we lack relevance but because we lack coordinated momentum.

This outcome is not inevitable.

EMS leaders, regulators, and policymakers have the power to design a smarter path forward—one that preserves patient safety, ensures clinician accountability, and invites innovation rather than fears it. To do so, we must modernize licensure laws, align data standards, and reduce friction between jurisdictions. We must design with intention so that EMS is not merely included but instrumental in shaping the healthcare systems of tomorrow.

Regulation should not be a barrier. It should be our launchpad.

When built collaboratively, it becomes a foundation for trust—a shared platform where innovation, safety, and accountability converge.

The path forward is not about choosing between rules and progress. It is about creating the rules that enable progress.

A Call for Equity and Performance

Performance consistency must be a core requirement for any AI deployment in EMS. Tools trained on narrow datasets may underperform in rural, frontier, or demographically diverse communities, undermining safety and fairness.

For example, AI triage models built on urban call data may misclassify risk in regions with different access patterns, language barriers, or community health needs. Speech-to-text systems, facial recognition tools, and predictive scoring engines have all shown reduced accuracy in underrepresented populations.

These are not just technical oversights—they are clinical risks.

EMS leaders have the opportunity and responsibility to demand technologies that perform reliably for all patients in all environments. This means pushing for transparency, validating tools across diverse settings, and ensuring that frontline feedback shapes implementation, not just post-deployment reaction.

Innovation without equity is not progress. It is exclusion.

The Cost of Doing Nothing

Innovation does not always fail because of bad ideas. Sometimes, it fails because leaders wait too long to act.

In 2007, Apple released the iPhone—a sleek touchscreen device with no physical keyboard and a heavy reliance on high-speed mobile data. At the time, BlackBerry dominated the business smartphone market, while Microsoft had its foothold in mobile enterprise platforms. Neither company was impressed.

BlackBerry executives dismissed the iPhone as a novelty. They insisted that business users would never give up a physical keyboard (fig. 4–3). They believed their proprietary, data-efficient systems were superior. Blackberry executives were not alone; Microsoft CEO Steve Ballmer even laughed on camera, mocking the iPhone's price point and limitations.

The executives were not stupid or inexperienced. They were anchored to their past success and blind to the shift that was already underway. Today, BlackBerry is a footnote in the smartphone market. Microsoft lost over a decade of opportunity and never recovered its mobile dominance. Meanwhile, the iPhone reshaped phones, photography, navigation, entertainment, and health.

This is what happens when innovative leaders underestimate disruptive change, and this is the crossroads EMS faces with AI.

It's tempting to laugh off AI-assisted dispatch, autonomous ambulances, or predictive triage algorithms. It's tempting to say, "That is not EMS. That is not what we do." That response is eerily similar to what the CEOs of Blackberry and Microsoft said about the iPhone.

As discussed throughout this book, AI is not coming—it has already arrived. It is reshaping every corner of health care: radiology leverages AI to read images

FIGURE 4–3. Once a leader in mobile communication, BlackBerry's rapid decline is a case study and warning to EMS leadership: technology evolution punishes complacency. Innovation must be continuous, not conditional.

with near-instant accuracy; intensive care units deploy predictive models to identify deterioration before it happens; sepsis detection tools scan vast data streams for early warning signs; and clinical systems now automate documentation, optimize staffing, and support diagnostic decision-making. EMS is not immune to this transformation—we are simply next. When integrated with clear intent, AI can elevate EMS across four critical domains: patient outcomes, operational efficiency, workforce morale, and public trust. This is not a gimmick or a tech fad. It is the beginning of a new era of intelligent medicine. EMS now stands at a decision point: We can defend the status quo, defer hard choices, or lead with clarity, urgency, and purpose.

This is not about abandoning our identity. It ensures that EMS remains a vital, visible, and forward-moving force in health care.

Innovation in EMS is constrained by budgets, policies, and legacy systems, but more importantly, innovation is shaped by what happens inside our minds. The most sophisticated strategy can fail if leaders fall into the same cognitive traps that stalled other industries. These psychological patterns are powerful, often invisible, and must be confronted with the same rigor we apply to protocol or policy. Before EMS can lead external transformation, we must lead internal clarity.

Cognitive Bias and Technology Integration in Emergency Medical Services

Bringing AI into EMS is not just a matter of systems and software—it is a matter of mindset.

Innovation often falters not because of a lack of vision or talent, but because of invisible barriers in thinking. Even the most capable EMS leaders can fall prey to unconscious biases that distort perception, slow adoption, and undermine progress. In a field defined by urgency, tradition, and high-stakes decision-making, these cognitive traps thrive under pressure, fatigue, and professional silos.

To lead wisely, we must think critically. To move forward, we must unlearn what no longer serves us. EMS leaders at every level must be given the tools—and the space—to grow their AI literacy. That includes seasoned professionals, including you.

Understanding these hidden psychological forces is not just helpful; it is essential. It is how we ensure that the future of EMS is not dictated by fear or default but by intentional leadership grounded in awareness, confidence, and collective purpose.

These biases are not signs of weakness but part of how the human brain manages complexity. But in moments of change, they can quietly distort judgment and slow innovation. The table below outlines the most common cognitive traps in leadership, including EMS leadership, and how to recognize and disrupt them (table 4–2). Mastering these patterns is not just an exercise in psychology. Leadership must guide the future of EMS.

TABLE 4–2. Cognitive Bias: How it Presents in EMS and How to Disrupt it

Cognitive Bias	What It Looks Like in EMS	Leadership Strategy to Disrupt It
Anchoring Bias	First impressions of AI—often outdated or negative—set the tone for all future decisions. Leaders stick to what they saw first.	Expose decision-makers to the latest tools and validated use cases. Break the first-exposure effect through current demos and case studies.
Availability Heuristic	A single anecdote ("This AI failed in Phoenix") overshadows extensive, successful data.	Ground decisions in patterns, not outliers. Share balanced evidence from EMS, nursing, and medicine.
Backfire Effect	When presented with evidence, skeptics double down on their original belief instead of reconsidering.	Focus on curiosity, not conversion. Use small pilots to build confidence rather than debate abstract benefits.
Black-or-White Fallacy	AI is seen as either a threat or a savior—no in-between. This polar thinking blocks practical, phased implementation.	Frame AI as a tool—neither magic nor menace. Ask: "Where does it help, and where doesn't it?"
Burden of Proof Bias	New tools face intense scrutiny, while outdated systems are accepted without evidence.	Apply equal rigor to legacy systems. Ask: "Would this pass today's standards if proposed now?"
Confirmation Bias	Leaders seek evidence that supports their current position—especially when evaluating vendors or pilots.	Assign a devil's advocate in every AI decision. Use structured evaluation criteria.
Curse of Knowledge	Experts explain AI in ways that alienate field personnel. Clinicians tune out due to inaccessible language.	Use plain language, analogies, and clinical context. Design education for the learner's level—not the expert's.
Dunning–Kruger Effect	Leaders overestimate their understanding of AI, resulting in poor planning or overconfidence in vendor claims.	Require baseline AI literacy in leadership roles. Normalize ongoing learning—even for senior staff.

Cognitive Bias	What It Looks Like in EMS	Leadership Strategy to Disrupt It
False Cause Fallacy	A system failure is wrongly attributed to "the new AI," even if unrelated. Innovation gets blamed unfairly.	Use root cause analysis. Separate correlation from causation before making changes or pulling tools.
Framing Effect	Clinicians reject or accept AI based on how it's introduced. "AI is replacing you" versus "AI supports you" triggers different reactions.	Frame AI as a teammate—not a replacement. Emphasize burden reduction and clinical enhancement.
Groupthink	Teams agree too quickly to avoid conflict—leading to fragile buy-in and overlooked risks.	Build psychological safety. Assign roles to challenge assumptions and introduce dissenting views.
Optimism Bias/ Negativity Bias	Some leaders ignore risks; others exaggerate them. Both miss the balanced reality of AI's limitations and value.	Use mixed-case studies and validation data. Set realistic expectations—not hype or fear.
Reactance Bias	Staff push back against AI simply because it feels imposed or externally driven.	Involve frontline clinicians early. Make them co-designers—not passive recipients—of innovation.
Sunk Cost Fallacy	Agencies cling to outdated platforms because "we already invested too much." Innovation stalls due to past spend.	Focus on future outcomes. Frame decisions in terms of opportunity cost and sustainability.
Anchoring Bias	First impressions of AI—often outdated or negative—set the tone for all future decisions. Leaders stick to what they saw first.	Expose decision-makers to the latest tools and validated use cases. Break the first-exposure effect through current demos and case studies.
Availability Heuristic	A single anecdote ("This AI failed in Phoenix") overshadows extensive, successful data.	Ground decisions in patterns, not outliers. Share balanced evidence from EMS, nursing, and medicine.

These are not abstract psychological concepts—they are the real reasons EMS systems stall, reject progress, or fall behind. *Innovation does not begin with new tools; it starts with clear thinking.*

EMS leaders who fail to confront cognitive bias will build systems that reflect it.

From Resistance to Readiness: Leading Cultural and Cognitive Change in Emergency Medical Services

Integrating AI into EMS is not just about software or equipment (in some ways, the software is easy!) but about trust, change, and leadership. The most common obstacles to innovation are not technical limitations but cultural hesitations and cognitive traps (fig. 4–4). To move EMS forward, leaders must do more than

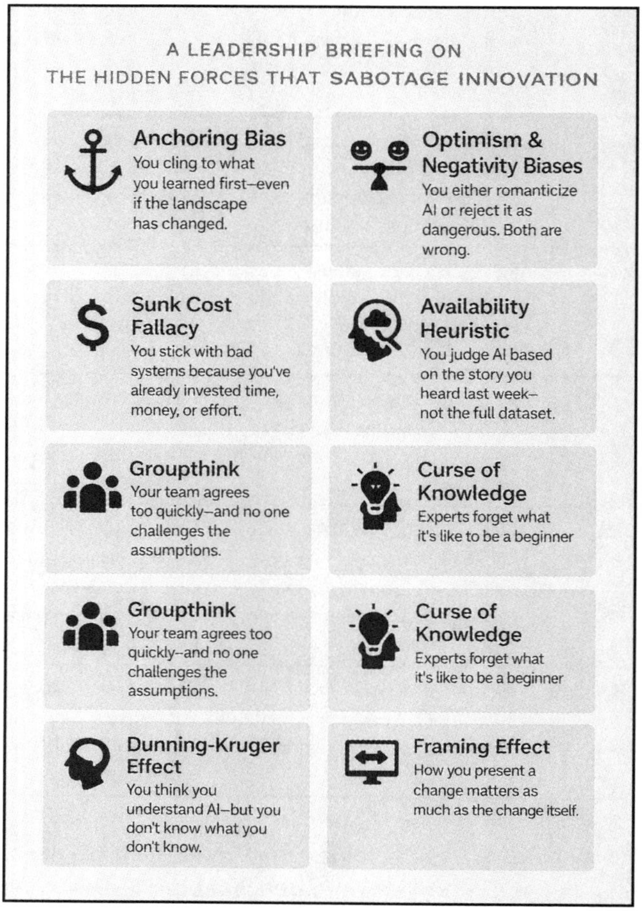

FIGURE 4–4. Biases such as anchoring, groupthink, and the Dunning–Kruger effect can distort how organizations evaluate innovation. Leaders must challenge assumptions before those biases harden into barriers to progress.

deploy tools. They must build environments where change is possible, resistance is addressed, and decisions are grounded in evidence, empathy, and professional integrity.

The following five strategy domains (table 4–3) provide a blueprint for overcoming resistance, aligning your agency with innovation, and ensuring that EMS leads, rather than lags, in healthcare transformation.

Recognizing these biases is not about assigning blame—it is about cultivating awareness. Great leaders do not avoid bias entirely; they learn to notice it, name it, and navigate through it. In an era shaped by AI, our thinking must be as adaptive and agile as the tools we choose to deploy.

This framework is not a call to rush adoption. It is a call to lead with clarity.

AI cannot—and should not—replace the human clinician. Its value lies in how it supports judgment, strengthens documentation, improves safety, and broadens access. However, that potential is only realized when EMS leaders build the conditions where AI can thrive—through trust, training, and thoughtful integration.

Innovation in EMS will not happen by accident. It will happen because leaders choose to act—with strategy, humility, and purpose.

If you want EMS to evolve, do not just change the tools. Change the culture.

TABLE 4–3. Strategic Leadership Framework

Strategic Focus	What EMS Leaders Must Do	Tactical Actions
Build Understanding	Most resistance to AI comes from lack of exposure, not bad intentions. AI literacy is essential for adoption.	• Integrate AI into onboarding, CE, and leadership training • Start simple: What is AI? How does it support EMS? • Share EMS-specific case studies
Create a Safe Culture	Innovation fails in fear-based environments. Psychological safety invites curiosity and critique.	• Assign a devil's advocate in new tech decisions • Use pre-mortems and post-mortems • Normalize discussion of failure as learning

(continues)

TABLE 4–3. Strategic Leadership Framework (*continued*)

Strategic Focus	What EMS Leaders Must Do	Tactical Actions
Lead with Clarity	How AI is framed shapes whether it is feared or embraced. Clarity builds trust.	• Avoid jargon • Frame AI as augmentation, not automation • Emphasize partnership, not replacement
Align Nationally	Fragmentation slows funding, research, and standard-setting. Unity amplifies EMS's voice.	• Support model rules and national datasets • Participate in cross-jurisdiction pilots and data-sharing • Contribute to national policy conversations
Implement Incrementally	Trying to do everything at once overwhelms teams and reinforces fear. Small wins build momentum.	• Pilot one workflow, module, or unit • Involve clinicians in iterative improvement • Share outcomes transparently
Confront Bias Directly	Hidden cognitive traps derail good decisions. Bias must be named and neutralized.	• Educate leaders on cognitive bias • Build systems to catch anchoring, groupthink, and sunk costs • Include clinicians in tool evaluations and post-mortems

Leading Through Resistance Is the New Competency

Leadership in EMS is no longer defined solely by clinical skill or operational efficiency—it is determined by our ability to guide teams through uncertainty, disruption, and change. AI will not wait for our permission. The healthcare system is evolving with or without us. And if EMS wants to remain relevant, trusted, and practical, we must match our clinical courage with strategic conviction. That means confronting resistance, rethinking our measures, and leading even when consensus is slow. Change will happen. The only question is whether EMS will shape or be shaped by it.

Key Terms and Definitions

- **Anchoring Bias:** The tendency to rely too heavily on the first piece of information encountered (the "anchor") when making decisions, often causing leaders to dismiss newer or more relevant data.
- **Burden of Proof Bias:** The tendency to demand extensive evidence for innovations while accepting legacy systems without equivalent scrutiny or evaluation.
- **Business Associate Agreement:** A legally binding document required under HIPAA that outlines each party's responsibilities in handling protected health information. EMS agencies must ensure that AI vendors meet these requirements to mitigate privacy and liability risks.
- **Change Management:** A structured approach to transitioning individuals, teams, and organizations from a current state to a desired future state, particularly in response to innovation or organizational transformation.
- **Cognitive Bias:** A systematic pattern of deviation from rational judgment that affects decision-making. In EMS leadership, cognitive biases can unconsciously influence resistance to new technologies or policy shifts.
- **Disruptive Innovation:** A term that describes new technologies or processes that significantly alter or replace existing systems, frequently resisted by incumbent organizations due to cultural or financial inertia.
- **Ethical AI:** AI designed, trained, and implemented in a manner that is transparent, equitable, explainable, and respectful of human rights. In EMS, this includes ensuring that algorithms do not perpetuate bias or inequity in patient care.
- **Framing Effect:** A cognitive bias where people's decisions are influenced more by how information is presented than by the actual content. In EMS, framing AI as a threat versus a support tool can significantly alter adoption behavior.
- **Groupthink:** A psychological phenomenon in which the desire for group consensus overrides realistic appraisal of alternatives, suppressing dissent and leading to flawed decision-making.
- **Innovation Gap:** The widening disconnect between the pace of technological development (e.g., AI, automation) and the ability of policy, regulation, or leadership structures to integrate and manage these advancements effectively.
- **Leadership Vacuum:** A condition in which key decisions are delayed, avoided, or made without strategic vision, often resulting in stagnation or lost opportunities for advancement.
- **Model Rules:** Uniform administrative policies or guidelines adopted across multiple jurisdictions to ensure consistent governance and

operational practices. In EMS, model rules for AI can support safe, coordinated innovation across state lines.

- **Pre-Mortem:** A strategic planning technique where a team assumes a project has failed and works backward to identify what could lead to failure—used in this chapter as a tool to avoid bias in innovation planning.
- **Reactance Bias:** A resistance to change driven by perceived threats to autonomy or control, common in EMS when external technologies or mandates are introduced without clinician input.
- **Regulatory Lag:** A condition in which policy frameworks and oversight mechanisms fail to keep pace with emerging technology, creating ambiguity, legal risk, or barriers to adoption.
- **Sunk Cost Fallacy:** A decision-making trap where past investments of time, money, or effort are used to justify continuing a failing or outdated project, rather than evaluating future value.
- **Waymo Moment:** A metaphor introduced in this chapter, referring to the disruptive entry of autonomous vehicles into traditional transport sectors—used here to illustrate EMS's current crossroads with AI and automation.

Reflection Questions

1. When did your organization last abandon a legacy system before it failed? What held you back?
2. Do your teams feel safe challenging leadership decisions, or do they default to an agreement? What structures have you put in place to break groupthink?
3. What cognitive bias have you caught yourself falling into during technology evaluations? What systems can you build to keep that from happening again?
4. Are you waiting for perfect conditions, or moving forward with the tools you have? What decision have you postponed that you need to revisit?
5. If an outsider evaluated your system today, would they describe your culture as adaptive or defensive? What would it take to change that?

Study Questions

1. Why must EMS leaders recognize that resistance to innovation often stems from cultural and structural systems, not technology itself?

2. Which of the following is NOT identified in the chapter as a systemic barrier to EMS innovation?

    a. Underfunding and misaligned incentives
    b. Fragmented governance
    c. Real-time triage algorithms
    d. Outdated infrastructure and training gaps

3. True / False: According to the chapter, systems—not tools—are the primary reason EMS innovation fails to scale.
4. Resistance in EMS often presents as ____________, making innovation harder to implement despite operational need.
5. What emotional and professional response did the suggestion of autonomous ambulances provoke within the EMS community?
6. What risks arise when AI tools used in EMS are trained on narrow or nonrepresentative datasets?
7. What metaphor does the chapter use to describe the danger of underestimating disruptive innovation in EMS?
8. What is one way traditional EMS regulations create obstacles for adopting AI tools?
9. List three cognitive biases that EMS leaders must confront when considering the adoption of AI.
10. What leadership strategies are recommended to shift EMS from resistance to readiness in the face of AI and emerging technologies?

CHAPTER SUMMARY

Lead Through the Barrier

AI is no longer emerging—it is here. However, EMS systems remain slowed not by a lack of tools but by structural barriers, regulatory inertia, cultural resistance, and unexamined bias.

This chapter examined four critical forces that block innovation:

- A system designed around transport, not transformation
- A culture shaped by tradition more than adaptation
- A regulatory framework built for yesterday's challenges, not today's technologies
- Leaders, often well-intentioned, but unaware of how deeply cognitive bias shapes their choices

These are not theoretical risks. The failure to adapt has operational consequences: longer response times, clinician burnout, missed funding, poor documentation, and declining public trust. EMS has a unique opportunity to ensure its voice and values shape the future of health care. Rather than being designed around, EMS can help lead design, bringing its field wisdom and adaptability to the center of innovation.

Yet the opportunity is real.

We can lead this transformation—but only if we lead it intentionally. That means understanding the difference between trend and strategy. It means building a culture that welcomes challenge and sees discomfort as the first signal of progress. It means developing national alignment so that innovation scales with integrity.

Leadership is no longer about defending the status quo. It is about preparing your agency and profession for what comes next.

Innovation in EMS is not a technical problem. It is a leadership mandate.

5

Health Care's Multibillion-Dollar Bet on Artificial Intelligence

LEARNING OBJECTIVES

After reading this chapter, the reader will be able to

- Explain the scope and trajectory of venture capital investment in AI across the healthcare sector, including historical shifts and emerging funding priorities.
- Evaluate the practical, legal, and strategic implications of ambient documentation, administrative automation, and predictive analytics in clinical care.
- Compare EMS's current posture toward AI adoption with broader trends in hospital systems, identifying key gaps in infrastructure, alignment, and leadership.
- Describe Moore's Law and the AI multiplier effect, and assess how exponential computing growth influences EMS readiness.
- Analyze barriers that have prevented EMS from participating in early AI development and identify actionable strategies to overcome them.
- Assess the impact of AI on healthcare cybersecurity infrastructure and outline why EMS must prioritize AI-enabled defenses.
- Articulate the significance of ImageTrend's private equity investment as a case study for the future visibility and viability of EMS technology partnerships.
- Formulate a leadership and policy strategy for integrating EMS into national and global AI frameworks, beyond pilot projects and individual vendors.

The Future Is Already Here

It was a routine appointment—December 2024. I was sitting in an exam room at my primary care physician's office, expecting nothing unusual. He is a capable, no-nonsense clinician in his mid-50s. When he walked in, he gestured casually to a small microphone clipped to his lapel.

"Hope it's OK," he said. "This helps me document the visit more quickly."

What followed was standard: a focused review of history, medications, recent travel, and a brief physical exam. He asked about work, made a few notes, and the visit ended without fanfare. But before returning to my car, my phone buzzed with a notification: *new visit summary available.* I opened my patient portal, and there it was—a fully written, neatly formatted clinical note. Every clinical detail of our encounter was captured and summarized with precision. It was clear, timely, and complete. It had been generated by an AI tool and reviewed by the physician. It worked precisely as intended.

Now, rewind to 2023.

I worked with colleagues to develop the first EMS-specific AI documentation tool that year—a HIPAA-compliant large language model trained on prehospital workflows. We believed it could reduce clinician burnout, improve documentation quality, and return valuable time to the field. The technology was sound, the use case compelling, and the early feedback from pilot sites encouraging, but we quickly ran into resistance.

It did not come from privacy regulators or national policy. The most significant barriers came from within the EMS ecosystem itself. Leading software vendors refused to collaborate or open their application programming interfaces (APIs), even though this type of interoperability is now standard in most of the broader healthcare sector. EMS agencies hesitated, unsure if AI could be trusted in clinical care. Medical directors questioned how they would review AI-generated narratives. And despite universal recognition that the tool could reduce liability, decrease frustration, and improve efficiency, no one wanted to pay for it.

By early 2024, we had a fully operational model explicitly designed for EMS, but the market was not ready. The whole story of that project—and what EMS documentation needs next—is explored in chapter 7.

I am no longer pursuing that project. I am proud of what we built. We sparked a conversation that had been missing for years. We raised awareness, demonstrated what was possible, and created a wave of demand. Every major ePCR vendor has now announced or released proprietary AI documentation tools, meeting expectations that barely existed when we started. In just a few months, AI-powered charting may become standard.

That was always the hope, but the experience left a lingering question: Why was EMS behind?

Why did it take so long for our profession to acknowledge a problem that clinicians had voiced for decades? Why did vendors resist the interoperability that the rest of health care has treated as a baseline for a decade? Why did leaders hesitate, knowing full well that the documentation burden was one of the most demoralizing parts of our profession?

And why do we keep waiting for others to go first?

This is not a story about a missed opportunity. It is a mirror. It reflects the broader pattern we see in EMS innovation: resistance masquerading as caution, fear presented as fiscal conservatism, and a culture that demands proof before it is willing to listen.

EMS did not fall behind in AI documentation because we lacked vision. We fell behind because we lacked alignment, infrastructure, and the confidence to act before consensus arrived.

This chapter is not about what went wrong. It is about where health care is going—and how much ground EMS must now make up because the rest of the industry is not waiting. Hospitals, physicians, and integrated health systems are already on their second and third-generation AI deployments. They are building tools not to test, but to scale.

If EMS wants to remain relevant in the era of intelligent care, we will need more than good ideas. We will need the courage—and the coordination—to act on them.

The Healthcare System Is Not Waiting

While EMS continues to debate whether AI is safe, trustworthy, or appropriate, the rest of health care has already moved on. Hospitals, health systems, and provider networks have tested AI and deployed it at scale.

In 2023, nearly 95% of U.S. healthcare organizations reported using AI in some capacity.[1] These are not fringe use cases. They span the entire care continuum—from predictive sepsis alerts in the intensive care unit to ambient documentation in outpatient clinics. The global AI healthcare market is expected to exceed $208 billion by 2030, reflecting widespread belief in AI as a core infrastructure, not a peripheral experiment.[2,3]

Here is what that looks like in practice:

- **Ambient documentation tools** automatically generate encounter notes during the visit, freeing clinicians from keyboards and drastically reducing burnout.
- **Intensive care unit monitoring systems** use predictive algorithms to detect subtle signs of deterioration hours before they become clinically evident.
- **Administrative AI platforms** now streamline scheduling, prior authorization, and denial mitigation, reducing human error while accelerating throughput.
- **Bed management and discharge planning systems** use AI to forecast length of stay and optimize patient flow.
- **Consumer-facing AI health platforms** from major technology companies have begun routing patients directly to AI-mediated triage, navigation, and care recommendations before any clinical encounter, effectively inserting a new upstream layer into the healthcare access pathway that EMS has historically occupied.

These are not pilots. They are standard practice, and they are being integrated into community hospitals and major academic centers.

The next wave is already past the hospital walls. Consumer technology companies have begun building AI-mediated front doors to the healthcare system, products that answer clinical questions, triage symptoms, and recommend next steps before a patient ever picks up a phone. These platforms are not designed by hospitals, and they do not include EMS. If they succeed at the scale their investors anticipate, the first clinical interaction for tens of millions of Americans will happen with a chatbot, not a 9-1-1 call. That changes what EMS must prove about its own relevance, and it shortens the timeline for making the case.

AI in health care is no longer a future investment—it is a present workforce multiplier.

And yet, EMS still struggles with the following:

- Paper-based quality assurance workflows.
- Manual entry of complex clinical narratives.
- Cultural discomfort with automation in clinical care.
- Health care has made AI necessary; EMS still treats it as negotiable.
- This is not just a gap in tools, it is a gap in mindset, and that gap is widening.

Artificial Intelligence Documentation: From Burnout to Breakthrough

One of the most widely adopted AI use cases in health care is ambient documentation. These tools capture real-time clinical conversations, summarize patient encounters, and generate structured notes, reducing administrative burden and giving clinicians more time.

Hospitals and outpatient practices are rapidly deploying these systems. Physicians using ambient documentation tools in primary care have reported a 50–70% reduction in charting time, fewer after-hours tasks, and improved note quality. These tools are not experimental but billing-compliant, HIPAA-secure, and embedded across major electronic health record platforms.

EMS, however, remains an outlier. While other sectors have embraced AI to mitigate documentation fatigue and reduce liability, most EMS clinicians still rely on manual, post-call charting under pressure. Structured fields dominate; narratives are often rushed. The result is growing frustration, delayed quality assurance, and missed opportunities to connect EMS to the broader healthcare data ecosystem.

The underlying issue is not technological—it is cultural and structural. Legacy systems, fragmented vendors, and hesitation around automation have slowed EMS adoption. Yet, ambient documentation may be one of the most urgent and impactful opportunities AI offers the field. We explore this transformation fully in chapter 7, where EMS documentation moves from burden to safeguard—and where AI becomes a vital partner in reclaiming narrative integrity, legal clarity, and professional confidence.

From MYCIN to Mayo—Five Decades of Artificial Intelligence Trust, Without Emergency Medical Services

AI did not arrive suddenly in the 2020s. It has been quietly reshaping health care for more than 50 years.

In the early 1970s, Stanford University researchers developed MYCIN, an early clinical decision support system designed to assist with diagnosing and treating bacterial infections.[4] While MYCIN never entered clinical deployment, it proved that computers could replicate expert reasoning and improve medical decision-making. It was the first demonstration that AI could have a place at the bedside. Chapter 2 provides a deeper overview of foundational AI definitions, model types, and the early history of systems like MYCIN.

That same decade, EMS was being born.

However, while EMS emerged from trauma transport and battlefield medicine, AI took root in academic hospitals and federally funded research centers. In 1979, the American Association for the Advancement of Artificial Intelligence[5] was founded to promote AI across disciplines. In the following decades, AI quietly expanded—from early cancer screening and robotic surgery to the systems that would eventually power modern electronic health records.[6]

Health care's confidence in AI was not built overnight. It was built through decades of infrastructure, validation, and interdisciplinary trust. Moore's Law—which predicts the doubling of computing power every two years—accelerated this trend, enabling tools that could process vast datasets and deliver real-time insights. Over time, these systems learned to follow rules, adapt, and improve independently.

When hospitals like the Mayo Clinic and Mount Sinai began deploying AI-driven sepsis alerts, image analysis platforms, and generative documentation tools, the groundwork had already been laid. The models had been trained, the partnerships had been formed, and the trust had been earned.

Meanwhile, EMS remained on the outside.

EMS was never part of the early AI conversation—not because it lacked relevance, but because it lacked institutional proximity. Without deep connections to research universities, federal funding pipelines, or interoperable data infrastructure, EMS missed the first 50 years of AI advancement.

That is not a failure. It is a starting point.

EMS now has the opportunity to leap forward—not by replicating the past but by compressing the learning curve. We must integrate with health care's existing trust networks, build academic and vendor partnerships, and apply the lessons already learned. EMS does not have five decades to figure this out, but it does have a choice: Catch up, or get left behind.

Health care's trust in AI was not built overnight. It was built over 50 years of tested infrastructure, interdisciplinary research, and exponential computing gains.

Moore's Law and the Artificial Intelligence Multiplier

The exponential growth of AI is not just a function of clever programming—it is powered by an economic and computational force that has quietly reshaped the digital world for decades: *Moore's Law.*

First articulated in 1965[7] by Intel cofounder Gordon Moore, the law predicted that the number of transistors on a microchip would double approximately every two years (fig. 5–1). For much of the past half-century, that prediction held. With each doubling, computing power increased, device costs dropped, and the accessibility of digital tools expanded. AI, in particular, has ridden this wave.

In recent years, AI has outpaced the original curve. Many researchers now estimate that computational power for AI tasks doubles every six months. This shift marks more than just faster processors. It reflects what some now call the *AI multiplier effect*. In this compounding phenomenon, each new generation of models improves speed, intelligence, nuance, and real-world applicability by building upon the training and architecture of prior systems.

This exponential layering of capability has the following implications:

- Tools that once transcribed clinical conversations now generate complete, structured narratives.
- Algorithms that once flagged sepsis now anticipate cascading organ failure.
- Systems that required extensive human oversight in 2020 can now outperform domain specialists in narrow but critical areas.

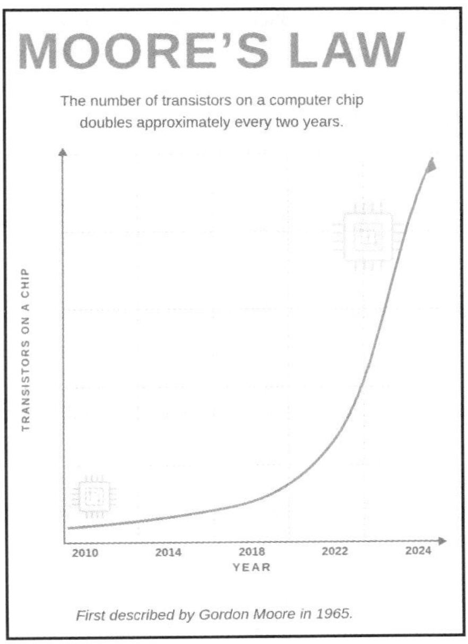

FIGURE 5–1. Moore's Law illustrates exponential growth in computing power. This doubling of capability every two years is propelling AI's rapid evolution. This is forcing EMS systems to adapt faster than traditional change cycles allow.

What makes this leap possible is not just silicon. It is the convergence of three exponential forces:

- **Computational abundance:** Processing power is increasing while costs continue to plummet.
- **Data abundance:** Healthcare systems generate structured and unstructured data at unprecedented volume.
- **Capital abundance:** Billions of public and private investment dollars accelerate AI research and deployment.

As Kat Davis[8] observed in her 2024 article "Moore's Law Accelerated," the marginal cost of computing is approaching zero, triggering a golden-ticket moment across industries. "Technologies that once seemed futuristic are suddenly within our grasp," she writes—and EMS must recognize the implications. In health care, AI is no longer a "nice-to-have" tool. It is becoming the default diagnostics, workflow, monitoring, and documentation infrastructure.

Why This Matters for Emergency Medical Services

In the past, EMS could afford to lag. We could wait for technology to stabilize, for costs to fall, for someone else to prove the concept. But exponential curves do not reward caution—they punish delay (fig. 5–2).

Each year that EMS postpones AI adoption, the performance gap significantly widens. Hospitals and insurers are optimizing care with predictive analytics, ambient documentation, and machine learning models embedded directly into clinical workflows. If EMS cannot interface with those systems—if we continue producing unstructured documentation, or lag in real-time data sharing—we risk exclusion from the next generation of care coordination.

This is not a technical mismatch. It is a strategic threat. We cannot slow the pace of AI, but we can accelerate our response. To remain relevant, EMS must

- build organizational readiness for AI-integrated workflows,
- invest in digital infrastructure that supports automation, interoperability, and real-time documentation, and
- adopt a mindset of cycles, not decades, where progress is measured in six-month iterations, not generational change.

The future will not arrive all at once. It is already coming—faster, smarter, and more integrated with each version release. EMS leadership must match that tempo. The cost of waiting is no longer just inefficiency; it is irrelevance.

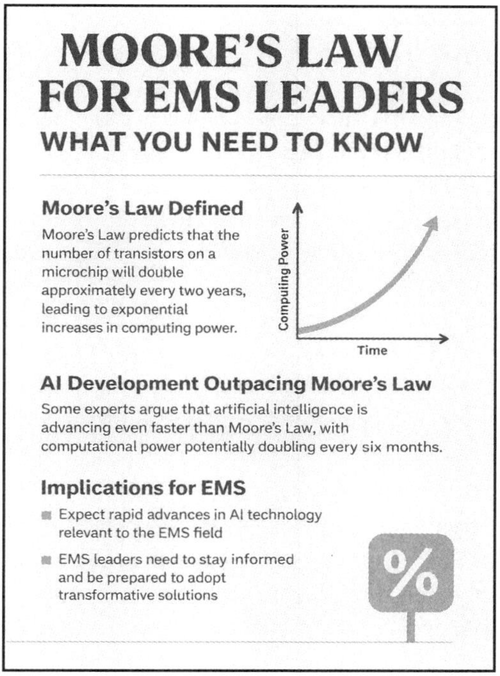

FIGURE 5–2. Moore's Law predicts exponential increases in computing power, which are now being outpaced by the development of AI itself. For EMS leaders, this means faster innovation cycles and the need for continuous learning to stay operationally relevant.

Follow the Money: Venture Capital and Strategic Investment

If you want to understand the future of health care, follow the money. And right now, that money is flooding into AI.

Since 2019, venture capital (VC) investment in healthcare AI has outpaced nearly every other sector of the tech economy. This is not a speculative bubble—it reflects a growing consensus that AI is the future of health care and the infrastructure of modern care delivery.

Healthcare AI funding has grown year after year throughout this period: over $7.2 billion in 2023 and more than $11 billion in 2024, with each subsequent year setting a new high.[9] One in every four healthcare venture dollars now goes to AI-enabled solutions. That capital is not chasing hype. It is chasing performance.

In 2024, U.S. healthcare companies received over $23 billion in VC, with nearly 30% of that total, $5.6 billion, channeled into AI-backed startups, a roughly threefold increase over the prior year.[10] That shift from exploration to execution has continued in every subsequent reporting period.

While earlier funding rounds emphasized biotechnology and drug discovery, the appetite for AI-powered diagnostics, decision support, and administrative automation continues accelerating. However, late-stage investors have become more selective, prioritizing proven tools over speculative disruption. As Silicon Valley Bank's 2025 report notes, "The flight to quality in health tech is in full force," with investors seeking clear, derisked value pathways before committing to major deals. As a sector, EMS must respond accordingly—with credible pilots, robust partnerships, and a unified voice capable of articulating our value in this evolving AI ecosystem.[11]

What Is Being Funded?

While early AI investment emphasized biotech and long-horizon research, today's funding is focused on tools delivering near-term operational value. In 2024, 60% of healthcare-AI funding targeted use cases like prior authorization, claims adjudication, clinical documentation, and revenue cycle optimization (Silicon Valley Bank, 2025).[12] These are not moonshot technologies—they are systems designed to reduce administrative burden, accelerate reimbursements, and stabilize clinical workflows under pressure.

This strategic redirection reflects investor preference for AI tools that face fewer regulatory barriers and offer immediate returns. The top investment categories include the following:

- **Administrative automation,** such as billing, claims management, and scheduling platforms
- **Clinical documentation,** including ambient AI scribes and voice-to-chart integrations within major electronic health record ecosystems
- **Diagnostic imaging,** where machine learning now rivals human interpretation in specific applications like radiology and CT scan triage
- **Predictive analytics,** powering real-time triage, early deterioration alerts, and dynamic resource allocation[13]

These investments are not speculative. They reflect a growing consensus that AI is the infrastructure of modern healthcare delivery, solving practical problems while also improving clinical accuracy and operational resilience. For EMS, this signals where partnerships and integration points must evolve if we expect to be seen as part of the broader healthcare solution.

In 2023, more than $1.5 billion[14] was invested in AI-powered diagnostic tools alone. Companies in this space are scaling rapidly—not because they promise something futuristic, but because they work now and can deliver

- faster cycle times,
- reduced clinician burden,
- higher diagnostic accuracy, and
- greater resilience for understaffed systems.[15]

The healthcare industry is not investing in AI because it looks impressive on a slide deck. It is investing because these tools solve real problems and generate measurable return on investment.

Artificial Intelligence and Cybersecurity: A Growing Imperative for Emergency Medical Services

AI is not just transforming clinical workflows—it is becoming essential infrastructure for protecting them. In February 2024, a ransomware attack on Change Healthcare disrupted nearly 40% of U.S. claims processing, exposing the healthcare system's vulnerability to cyber threats.[16] In response, healthcare organizations accelerated their investment in AI-powered cybersecurity tools, shifting from passive defenses to intelligent, adaptive threat detection.

According to IBM's 2024 *Cost of a Data Breach* report,[17] healthcare organizations using AI and automation could detect and contain cyber incidents nearly 100 days faster than those that did not. On average, they saved over $2.2 million per breach—a 45% improvement over systems without AI-powered security layers.

These savings are not theoretical. AI-enabled cybersecurity tools now play a role in

- real-time threat detection and prevention,
- rapid incident investigation and containment, and
- data protection across cloud-based EMS documentation platforms.

As EMS systems become more connected and data-rich, cybersecurity is no longer a hospital problem but an EMS problem. AI is the only tool that can scale fast enough to monitor, defend, and adapt to emerging threats across thousands of nodes in real time. If EMS agencies want to secure their infrastructure and maintain public trust, AI-enabled cyber defense must be part of the modernization plan.

A Rare Glimpse into Venture Capital and Emergency Medical Services: The ImageTrend Inflection Point

For years, EMS has remained outside significant VC and private equity conversations. The reasons are well known: fragmentation, a relatively small market size, and a profession historically cautious about adopting unproven technologies. But in 2023, that narrative began to shift.

ImageTrend, Inc., a leading EMS and healthcare software company, received a strategic growth investment from Welsh, Carson, Anderson & Stowe[18], a prominent private equity firm specializing in technology and health care. The investment was designed to accelerate innovation, expand reach, and strengthen the company's data and analytics capabilities.

Within months, the ripple effects were apparent:

- **AI Assist,** a voice-to-field documentation tool designed to reduce narrative fatigue and improve data quality, was launched.
- **Biospatial,** one of the country's most advanced EMS data intelligence platforms, was acquired and integrated.
- **ImageTrend** began aligning its core identity around analytics, intelligence, and interoperable system integration.

This inflection point marked a shift for ImageTrend and perhaps for EMS.

Why It Matters

A few years earlier, EMS leaders hesitated to explore these technologies. Many cited privacy concerns, clinician skepticism, or regulatory uncertainty. Some vendors resisted integration or refused to open their APIs, practices unheard of in broader health care IT. EMS agencies knew the tools could save time and improve quality, but few were willing to be the first to implement them. No one wanted to pay for what felt unfamiliar.

Yet once capital entered the equation, alignment followed. When a significant investment signaled market confidence, the technologies considered too experimental suddenly became fundable, deployable, and worth serious attention.

This is not a declaration of victory. It is a proof point.

Strategic investment creates clarity, and clarity accelerates adoption. EMS is becoming visible to investors, innovators, and national strategy leaders. But visibility is not enough. We must act.

The Global Picture: Strategic Investment and National Alignment

AI is not just reshaping health care within the United States—it is transforming entire economies. Across the globe, countries are positioning AI as a cornerstone of national development. And the sectors receiving the highest investment? According to the World Economic Forum and the Organization for Economic Co-operation and Development, it's health care and autonomous vehicles.

That intersection is not theoretical. It is EMS.

The Scale of Investment

Private-sector AI investment has exceeded $290 billion in the United States over the last five years. As the World Economic Forum recently noted:

> *In the U.S., a fervent gold rush in private AI investment has catalyzed transformative developments across key sectors such as autonomous vehicles, health care, and IT infrastructure. VC investments in AI totaled $290 billion over the last 5 years. In the U.S., optimistic projections suggest that AI could boost annual GDP growth by 0.5 to 1.5% over the next decade. That is $1.2 trillion to $3.8 trillion in real terms.*
>
> —World Economic Forum[19]

Health care and mobility/autonomous vehicles dominate U.S. VC investment in AI, accounting for nearly $100 billion (fig. 5–3). These are not isolated trends; they are strategic priorities driving national innovation. EMS stands at its intersection.

Yet EMS remains largely absent from these conversations. That absence is not just symbolic—it is structural. While AI transforms logistics and patient care in adjacent fields, EMS still lacks the policy alignment, research participation, and investment visibility to capture its rightful share of this transformation.

This is not hypothetical. These investments actively reshape diagnostics, logistics, emergency transport, and real-time clinical decision-making.

In Sri Lanka, that reshaping is already visible. Despite economic collapse and healthcare system strain in 2022–2023, Sri Lanka's national EMS system launched a low-cost digital transformation using cloud-native platforms and multilingual mobile documentation. In partnership with regional developers, they deployed an AI-enabled tool that helped field clinicians document patient

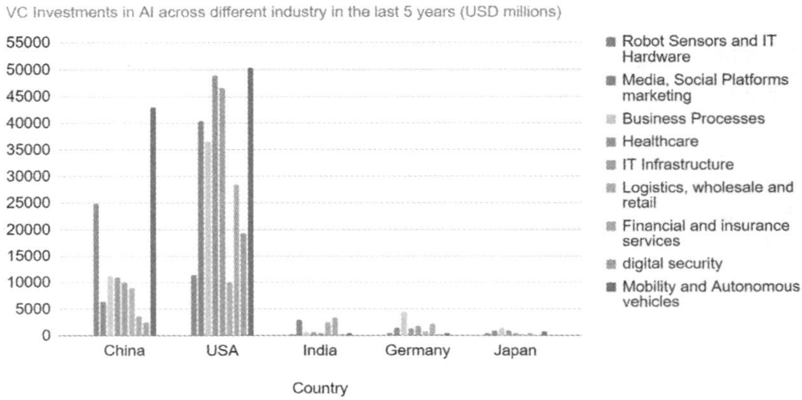

FIGURE 5–3. Over the past five years, the United States and China have led the way in AI investment, particularly in health care, mobility, and financial services. EMS leaders must recognize these global trends as indicators of where innovation will next influence prehospital care.[20]

care in real time, even in rural or low-bandwidth settings. The system translated symptoms across Sinhala, Tamil, and English; generated structured data compliant with national reporting standards; and integrated with hospital triage systems upon arrival. It was not expensive, but it was aligned. Leadership, simplicity, and interoperability—not budget—enabled progress. The implication is clear: Innovation does not require a massive budget, it requires coordinated will. If Sri Lanka can leapfrog outdated infrastructure and deploy intelligent EMS documentation, then the United States has no excuse for lagging behind.

Globally, momentum is accelerating:

- China is investing over $60 billion in AI for health care, backed by national plans like *Made in China 2025*.
- Japan uses AI to address its aging population, especially in remote monitoring and healthcare automation.
- The European Union is crafting cohesive AI policy frameworks that balance innovation with ethics.

Emergency Medical Services: Positioned, but Disconnected

EMS operates at the front line of healthcare delivery and mobile automation—the nexus of these two investment priorities. No other discipline handles point-of-care diagnostics, emergency transport, and system-wide navigation under urgent conditions while generating real-time health data. Yet, EMS remains absent mainly from AI research initiatives, policy frameworks, and investment pipelines.

This is not due to a lack of relevance. It is due to a lack of coordinated engagement.

A National Imperative

EMS must act now—not just to adopt AI tools, but to define our role in this transformation. That means engaging in national strategy conversations, launching high-visibility pilot projects, and partnering with health care, research, and technology leaders to demonstrate value.

Leadership is about recognizing the change and claiming our place in it.

Why Emergency Medical Services Can Not Afford to Stay Silent

Today, EMS remains largely absent from national AI strategies and investment priorities. We are underrepresented in venture funding, often excluded from cross-sector research collaborations, and overlooked in regulatory modernization efforts. This is not because EMS lacks value—it is because, as a profession, we have not positioned ourselves to be visible, fundable, or strategically aligned with AI's national momentum.

That silence has consequences:

- We have not *unified our voices* across states, agencies, or national associations to define what AI-enabled EMS should be.
- We have *deprioritized innovation*, often favoring legacy systems or "wait-and-see" approaches over bold experimentation.
- We have *failed to articulate* EMS's critical role in data creation, system navigation, community health, and emergency preparedness.

If EMS does not actively align with national and global AI priorities, we will not simply fall behind—we will be designed around. The future workflows will be built by those present in the conversation. Those absent may find themselves subject to tools, protocols, or mandates that do not reflect the complexity or urgency of field-based emergency care.

The Good News: We Do Not Need Billions

EMS does not need to secure massive VC rounds to catch up. We need the following:

- **Coordinated leadership** across agencies, states, and associations
- **Credible pilot programs** that demonstrate real-world value and interoperability
- **Strategic partnerships** with universities, hospital systems, and vendors already in the AI space
- **An internal culture** that rewards innovation, reduces fear of change, and creates clear pathways for responsible adoption

The question is no longer whether AI will transform health care—it already has. The question is whether EMS will be part of that transformation or a relic of the previous era.

Why Emergency Medical Services Must Align with National Artificial Intelligence Strategy

Admiring AI's success in health care is not enough. EMS must study its strategy and adopt one of its own. The sectors benefiting most from AI investment do not necessarily have the most significant clinical need. They are the ones that demonstrate alignment.

Funding, research, and policy support flow toward systems that present a unified national strategy, show regional interoperability, and offer a clear return on investment. These sectors speak the language of innovation, implementation, and impact. At present, EMS does not meet those criteria consistently—but it can.

The EMS Compact (known as REPLICA) offers a successful precedent. By aligning states under model legislation and shared governance, the Compact has made interstate EMS licensure mobility possible without sacrificing safety or accountability. The same approach is needed for AI.

If EMS wants to be recognized as an essential partner in AI-enabled health care, we must move beyond fragmented pilot programs and build national coherence. That means developing documentation standards that align with AI tools, fostering academic and public-private partnerships, and articulating a clear vision for what AI-enhanced EMS can achieve.

This is not about chasing hype or competing with hospitals on scale. It is about matching health care's commitment to clarity, coordination, and credibility. Sectors that succeed in AI adoption do not wait for permission. They organize, experiment, measure, and lead.

Now is the time for EMS to do the same.

Why Emergency Medical Services Is Uniquely Positioned—But Unprepared

EMS sits at a rare and powerful intersection. Few sectors span as many domains as EMS, simultaneously involving healthcare delivery, public safety, logistics, emergency management, and, increasingly, data-driven community health (fig. 5–4). From triaging a 9-1-1 call to navigating dynamic field conditions, EMS clinicians provide mobile, real-time, high-stakes care across every geography and demographic.

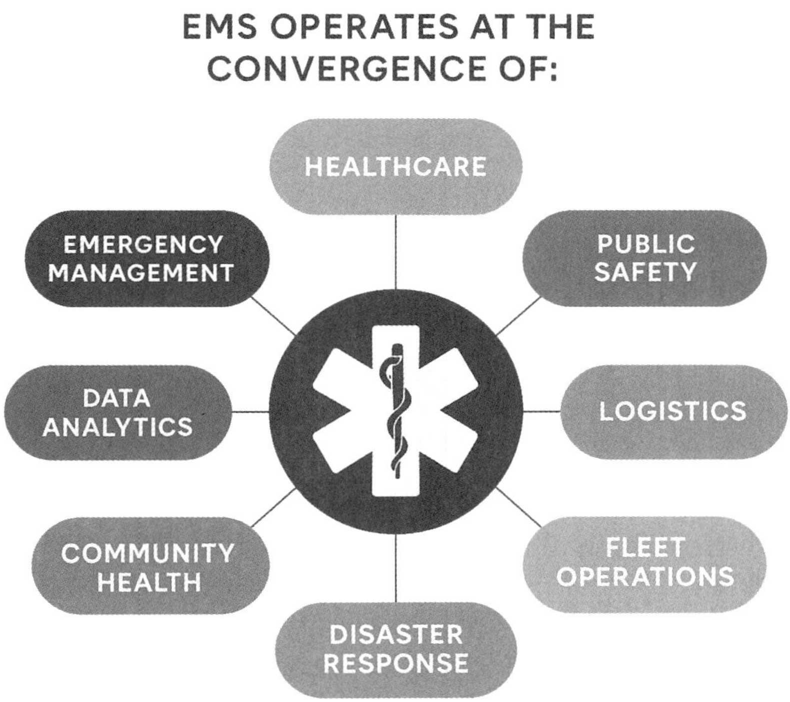

FIGURE 5–4. EMS stands at the center of health care, logistics, data analytics, and disaster response. AI technologies can unify these functions, improving coordination and decision-making across domains.

No other clinical profession is tasked with so many roles at once: diagnostics, transport, crisis response, resource allocation, and systems navigation. EMS connects the home to the hospital, the moment of need to the broader continuum of care. If AI is meant to enhance mobile, time-sensitive, and information-rich health care, then EMS is not just a relevant player but a critical one.

Yet despite this unique positioning, EMS remains unprepared mainly for AI adoption at scale.

We lack shared documentation standards that align with emerging AI tools, interoperable systems that can integrate with hospital and insurer platforms, coordinated academic partnerships, multiagency pilot programs, and national-level research initiatives to define what AI in EMS should look like, and, perhaps most importantly, alignment—a unified voice to articulate our values, needs, and goals.

This is not a reflection of failure. It is a reflection of fragmentation. For decades, EMS has been built around local governance, variable funding streams, and operational independence. That autonomy has many benefits, but it creates real costs in the age of AI. Innovation depends on scale, structure, and shared vision. Without them, EMS will continue to be overlooked by funders, regulators, and developers—not by malice, but by design.

EMS has always been relevant—what we need now is the readiness and alignment to meet this moment of transformation with confidence and coordination.

That can change, but only if EMS leaders choose to organize, invest, and lead with strategic clarity.

This Is Our Moment to Act

AI is not just transforming health care—it is redefining its infrastructure, accelerating its pace, and reshaping expectations. Major health systems are redesigning clinical workflows around AI. Federal agencies are investing in infrastructure to support algorithmic decision-making and digital coordination. Private equity firms are targeting scalable, software-enabled care. These forces are not hypothetical—they are already determining who leads, who lags, and who gets left behind.

EMS is at a decision point. We can either position ourselves as the frontline intelligence layer of emergency health care or be treated as an operational afterthought, folded into someone else's design. That choice will not be made for us; it will be made by us.

The good news is that we do not need to match hospital systems' funding or institutional heft to claim our place. What we need is focus. EMS must speak with a unified voice, pilot credible solutions, form strategic partnerships, and demonstrate measurable impact. Most importantly, we must shift from reactive to proactive framing AI not as an external pressure but as an internal opportunity to reimagine how we deliver care.

This is a rare inflection point. For once, the technology, capital, and clinical needs are aligned. But if we remain uncoordinated—if we wait for perfect conditions or a national mandate—EMS will be built around and not included.

Innovation is not a spectator sport. The healthcare system is evolving with or without us.

Now is the moment for EMS to decide what role we want to play in the future of care.

From Investment to Implementation: Emergency Medical Service's Next Mission

The message is clear: AI reshapes health care's infrastructure, workflow, and economics. But the conversation must now move beyond investment headlines and policy papers. What comes next is implementation, where EMS leadership must rise.

It is not enough to know that AI is being funded. EMS must learn how to

- evaluate tools beyond vendor hype,
- set expectations around workforce training and change management, and
- align AI implementation with quality metrics, operational key performance indicators, and clinical goals.

Hospitals are no longer waiting to understand AI—they are shaping how it is used. EMS must do the same. That means asking hard questions such as

- How will we define "success" in an AI pilot?
- Who will be responsible for oversight, bias detection, and system validation?
- How will we protect our clinicians and patients by adopting learning and evolving tools?

The future of EMS will not be decided by how much funding AI receives. It will be decided by how effectively we implement it.

Key Terms and Definitions

- **Ambient Documentation:** The use of AI-powered tools to capture, transcribe, and structure clinical conversations in real time, enabling documentation with minimal manual input.
- **AI Multiplier Effect:** A compounding phenomenon where each generation of AI models builds on the capabilities of prior systems, accelerating gains in accuracy, speed, and contextual awareness.
- **Capital Abundance:** The influx of public and private investment into AI technologies accelerates development and deployment across healthcare sectors.
- **Clinical AI:** AI applications are used directly in patient care, such as diagnostic decision support, triage tools, and treatment recommendations.
- **Cybersecurity AI:** AI-enabled systems that detect, prevent, and respond to cyber threats in real time, particularly critical in data-rich, digitally integrated healthcare environments.
- **Interoperability:** The ability of different IT systems, software, and tools to communicate, exchange, and interpret shared data effectively across healthcare settings.
- **Moore's Law:** The observation that the number of transistors on a microchip doubles approximately every two years, historically driving exponential growth in computing power.
- **Prior Authorization:** A cost-control process requiring providers to obtain approval before delivering certain healthcare services. AI tools now automate portions of this workflow.
- **REPLICA** (Recognition of EMS Personnel Licensure Interstate CompAct): A model legislation that enables interstate practice of EMS clinicians and serves as a governance example for broader policy alignment.
- **Venture Capital (VC):** Private equity investment provided to early-stage, high-potential startups. In health care, VC is now disproportionately focused on AI-based innovations.

Reflection Questions

1. Are we, as a profession, aligning our AI strategy with national and global healthcare trends—or waiting to be pulled along by them? What are we doing locally or organizationally to stay ahead of the curve?

2. How can EMS demonstrate its unique value at the convergence of health care, logistics, and public safety to attract attention, funding, or partnerships? Are we clearly articulating our role as both data creators and frontline clinicians?
3. What lessons can EMS learn from the healthcare industry's success in securing VC and national policy alignment? How can we replicate that to reflect the operational realities of EMS?
4. What stands in the way of greater AI adoption in EMS—technology, funding, policy, or culture? And which of those barriers are self-imposed?
5. If AI is expected to reduce burnout, improve documentation, and optimize decisions, can we afford not to invest in it—at least strategically and educationally?

Study Questions

1. According to this chapter, why is EMS at risk of being excluded from next-generation healthcare innovation?
2. Which of the following best describes the primary driver behind AI's rapid adoption in hospitals and health systems?

    a. Public resistance to manual care
    b. Pressure from EMS unions
    c. Demonstrated operational return on investment and clinical performance
    d. Mandatory federal legislation

3. True / False: Ambient documentation tools are experimental technologies currently in limited testing across hospitals.
4. What historical factors contributed to EMS being left out of the early decades of healthcare AI development?
5. How does Moore's Law—and the acceleration beyond it—impact the pace of AI advancement in health care?
6. What are three specific healthcare use cases for AI that are already deployed at scale outside of EMS?
7. What warning does the chapter offer through the metaphor of the "BlackBerry versus iPhone" moment? How does it apply to EMS?
8. According to the chapter, which two sectors have received the highest AI investment globally—and why is this especially relevant to EMS?
9. List two reasons why EMS remains unprepared to scale AI adoption, despite its relevance.
10. What leadership actions must EMS agencies take to align with national AI strategies and avoid being "designed around" in future care systems?

CHAPTER SUMMARY

Health Care's Multibillion-Dollar Bet on Artificial Intelligence

AI is no longer speculative—it is the new operating system of modern health care. With over $11 billion invested in 2024 alone, AI has moved from proof-of-concept to full-scale deployment across hospitals, health systems, and payer organizations. This chapter explores the economic, historical, and technological forces driving health care's rapid AI adoption and identifies why EMS has been largely absent from this transformation.

AI is already solving critical pain points in healthcare operations, from ambient documentation and predictive triage to administrative automation and cybersecurity. EMS, by contrast, remains hindered by internal resistance, vendor fragmentation, and a lack of national alignment. While health care has treated AI as core infrastructure, EMS continues to treat it as optional. This divergence is not only delaying progress—it is placing EMS at risk of being excluded from next-generation care coordination altogether.

The chapter traces the 50-year evolution of trust in AI from MYCIN to the Mayo Clinic, explains Moore's Law and the accelerating AI multiplier, and examines the capital shift toward tools with near-term operational return on investment. It highlights VC trends, cybersecurity imperatives, and EMS's emerging role—exemplified by ImageTrend's private equity investment—as signals of urgency and opportunity.

Most importantly, the chapter issues a call to leadership. EMS does not need to match hospital funding, but must match hospital focus. Strategic alignment, credible pilot programs, interoperable documentation systems, and national policy engagement are now essential. AI will not wait for EMS to be ready. If EMS leaders want to preserve relevance, they must now act to adopt tools and define how EMS will lead in the age of intelligent care.

6

The President of the United States, Policy, and the National Push for Artificial Intelligence in Health Care

LEARNING OBJECTIVES

After reading this chapter, the reader will be able to

- Explain the current federal landscape for AI in health care, including recent executive orders and major agency initiatives from the Biden and Trump administrations.
- Compare the AI governance frameworks of major healthcare organizations (e.g., AMA, ANA, ACP, AAFP) and articulate how their values align with the EMS mission.
- Assess the risks of EMS being excluded from national AI policy development and identify strategies to ensure representation and input.
- Analyze how partnerships between companies like Microsoft and Epic have created scalable models for responsible AI integration, and evaluate the lessons EMS can draw from them.
- Apply the AMA's three essential clinical AI questions: Does it work? Does it work for my patients? Does it improve outcomes?
- Formulate key components of an EMS-specific AI framework, including national collaboration, clinician education, and ethical safeguards.
- Evaluate the strategic need for open architecture in EMS software platforms, and explain how it enables interoperability and innovation.
- Anticipate the consequences of inaction, including loss of professional agency, misaligned systems, and external control over prehospital care technology.

Prologue: From Press Releases to Patients

In late 2023, President Biden announced sweeping actions on AI. His executive order did not just signal intent—it launched a federal race to develop AI policies across every major agency. New standards were issued to evaluate AI tools in clinical care at the Department of Health and Human Services. At the U.S. Department of Transportation, officials were ordered to develop a policy for AI in autonomous vehicles. At the Department of Defense, procurement officers were given new directives to ensure ethical guardrails for AI on the battlefield.

The White House's implications were clear: AI can improve health outcomes, reduce costs, and reduce disparities.

The intent is there, and the direction is clear. AI is not a fringe idea anymore—it is a full-blown national priority. AI is framed as the engine of future performance across government, health care, logistics, infrastructure, and defense.

But this was not just a Democratic initiative.

Four years earlier, in 2019, President Trump launched the Maintaining American Leadership in Artificial Intelligence initiative, directing federal agencies to invest in AI research and development, expand data infrastructure access, and establish safety and ethics standards. Then, in January 2025, President Trump returned to the conversation with a new directive: Executive Order 14179—"Removing Barriers to American Leadership in Artificial Intelligence." Shortly after revoking Biden's 2023 executive order, this new order directed all federal agencies to reevaluate AI policies that might impose regulatory constraints or ideological bias. It also tasked top science, economic, and national security officials with crafting America's AI Action Plan[1] to reinforce America's global AI leadership.

Shortly afterward, the Trump administration announced support for the Stargate Project, a multibillion-dollar private-sector initiative led by OpenAI, Oracle, and SoftBank to build domestic AI compute infrastructure on U.S. soil. The administration framed it as strategic investment in secure, domestic AI development unconstrained by foreign dependencies.

As the baton passed between administrations, it changed form. While Biden's order emphasized safeguards and equity, Trump's vision prioritized deregulation and economic acceleration; however, both presidents clarified that AI is a national priority through their distinct lenses.

Despite being a national priority, other sectors have been named in these sweeping initiatives—transportation, defense, cybersecurity, and education. EMS remains conspicuously absent from every White House directive, federal framework, and advisory council.

It is not because EMS is not essential. This absence is not due to a lack of importance—EMS is integral to health care, public safety, logistics, transportation, disaster response, and, increasingly, community health. EMS occupies a vital intersection within the broader healthcare system.

The absence of EMS from national AI discussions is a multifaceted issue. While a lack of unity within the EMS community may play a role, other factors, such as visibility, historical context, and competing priorities, also contribute to this oversight. To ensure that EMS is represented in future AI initiatives, the EMS industry must unite. It is essential to present a cohesive front to national leadership, moving beyond fragmented delivery models to advocate for EMS as a singular industry with a vital role in health care, national security, and disaster management. By doing so, we can effectively highlight our critical contributions and ensure that EMS is included in the national conversation on AI.

The federal government is moving fast. If EMS does not speak with a collective voice, align with national policy, and establish our own frameworks for AI integration, we will be excluded by default, and in that silence, critical decisions about the future of emergency care, documentation, diagnostics, and deployment will be made without us.

The key to having a voice is positioning. It is time for EMS to actively engage in the national conversation on AI, not as spectators but as contributors, leaders, and essential stakeholders in the future of health care. Let us be clear: AI is not a distant or abstract concept, but rather a present imperative.

If we do not act, others will define that future for us.

Connecting Emergency Medical Services to the Federal Strategy Table

AI policy does not just evolve within professional associations—it is shaped at the highest levels of national governance. Since the early 2000s, every U.S. president has issued technology initiatives and executive orders influencing which sectors receive funding, regulatory clearance, and research prioritization.

In 2023, President Biden's Executive Order on Safe, Secure, and Trustworthy AI established a government-wide directive to integrate AI into health care, transportation, education, and defense—focusing on equity, ethics, and data privacy. In early 2025, President Trump reversed that order and issued a new one—Executive Order 14179—directing agencies to accelerate AI development and remove regulatory barriers, launching initiatives like the Stargate Project to build domestic AI infrastructure.

Despite these divergent approaches, one thing is constant: AI has become a strategic national asset. These executive orders have guided agency funding, contract procurement, and program priorities for years.

But nowhere in these strategies is EMS mentioned.

This silence is not because EMS lacks value—it is because EMS has not claimed its position. If EMS wants to be part of national AI implementation, it must engage with agencies like the Office of Science and Technology Policy, the U.S. Artificial Intelligence Safety Institute at the National Institute of Standards and Technology, the Department of Health and Human Services, and the Department of Homeland Security.

Federal inclusion is not symbolic. It determines whether EMS receives grant funding, participates in pilot programs, and shapes standards that impact every clinician's daily tools. EMS leaders must move quickly to respond to AI and ensure AI is built with EMS needs in mind.

If we are not present in the planning, we will be subject to the results of someone else's blueprint.

National Frameworks for Artificial Intelligence in Health Care

AI policy in American health care did not evolve by accident. It was deliberately built over many years by national medical associations, research organizations, and federal agencies that understood what was at stake. They did not wait for disruption; they planned for it.

Since 2020, leading medical organizations like the American Medical Association (AMA), American Nursing Association (ANA), American College of Physicians (ACP), and American Academy of Family Physicians (AAFP) have published comprehensive policy frameworks guiding AI adoption. These documents emphasize ethics, transparency, equity, and practical integration and provide precisely the kind of leadership EMS has yet to deliver.

These were not academic whitepapers buried in journals. They were bold declarations: *AI must serve clinicians, not replace them. AI must reduce disparities—not amplify them. AI must integrate with workflow—not complicate it.*

American Medical Association: From Augmented Intelligence to Trustworthy Artificial Intelligence

In 2019, the AMA formally shifted its terminology from "artificial intelligence" to "augmented intelligence"[2]—a signal that the goal was not to replace physicians but to enhance their effectiveness. Their framework was built on the Quadruple Aim model of improving outcomes, reducing costs, enhancing patient experience, and supporting clinician well-being. They clarified that AI must relieve the administrative burden, not add to it.

Their 2023 Trustworthy Augmented Intelligence in Healthcare initiative[3] added depth, outlining specific questions clinicians should ask of AI vendors and urging organizations to align AI innovation with medical ethics, clinical validation, and continuous improvement cycles (fig. 6–1).

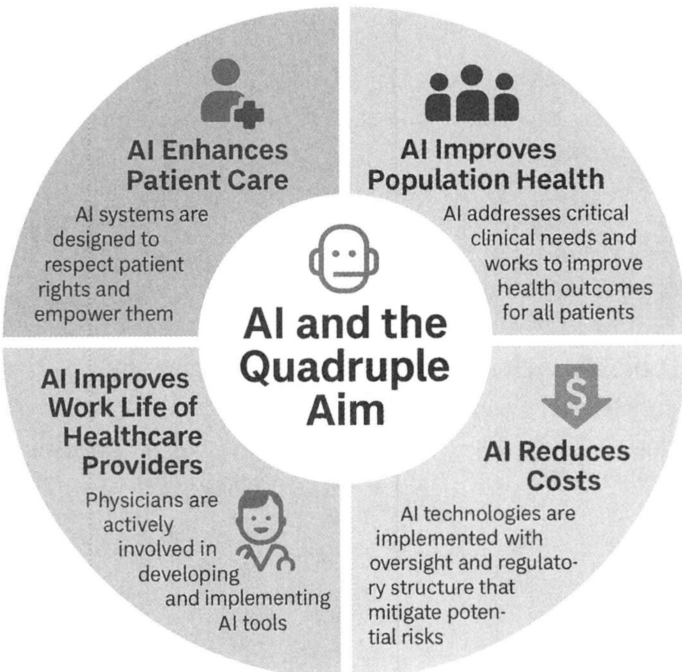

FIGURE 6–1. AI supports the Quadruple Aim by enhancing patient care, improving population health, reducing costs, and improving the work-life balance for clinicians. These same goals must guide the implementation of AI within EMS.

American Nursing Association: Nursing Ethics Meets the Algorithm

The ANA was not far behind. In 2022, it published "The Ethical Use of Artificial Intelligence in Nursing Practice,"[4] which put patient privacy, informed consent, and health equity at the center of AI policy. Their position was clear: Nurses must help design, deploy, and govern AI—not simply be expected to comply.

The ANA also emphasized workforce preparation. If AI is to succeed in clinical settings, clinicians must be trained to evaluate it, question it, and intervene when necessary. In this way, AI literacy becomes an extension of professional competence.

American College of Physicians: Artificial Intelligence Should Reduce Disparities, Not Reinforce Them

In 2024, the ACP released a significant position paper titled "Artificial Intelligence in the Provision of Health Care." Its headline was "AI must support—not supplant—clinical decision-making." But the document went further.

It demanded that AI models be validated using real-world patient data from diverse demographics. It called for multisector collaboration—federal agencies, private companies, academia, and nonprofits—to monitor for algorithmic bias and ensure transparency.

ACP's paper was not just about theory. It laid out ten clear policy recommendations—covering clinician education, data privacy, oversight mechanisms, adverse event reporting, and environmental sustainability.

In short, it was the kind of roadmap EMS needs now.

American Academy of Family Physicians: Real Tools for Real Problems

As far back as 2021, the AAFP was already field-testing AI assistants to reduce the documentation burden.[5] Their research-backed position emphasized the practical value of AI—less time typing and more time with patients. Their focus was not the hype—it was the clinician experience. And that is exactly the lens EMS must adopt.

These frameworks were not just about vision—they were about discipline. Each organization asked the hard questions, took ownership of its role in AI integration, and did not wait for federal mandates. They led.

Meanwhile, EMS as an industry has yet to produce a single national policy framework on AI. There is still no formal ethics guidance from a profession-wide body, no unified strategy for documentation tools, and no professional consensus on clinician–AI interaction.

That is not a gap—it's a liability.

Now is the time for EMS to catch up with the technology and the governance surrounding it.

Translating Principles into Practice

Vision is meaningless without execution. Every national medical framework discussed, including AMA, ANA, ACP, and AAFP, goes beyond lofty ideals. They offer structure, accountability, and a clear delineation of who must do what for AI to be safe, effective, and equitable.

And here's the key insight: These frameworks were not written by monolithic professionals with uniform job descriptions. Nurses work in environments as varied as school clinics, cardiac surgical suites, community health programs, occupational medicine, and home-based care. Physicians span family practice, emergency departments, specialty clinics, and inpatient units. Yet despite this extraordinary diversity, each profession came together, not based on where they work, but on who they are.

They are defined not by their setting, but by their professional identity.

Under that identity, they created unified national frameworks to guide AI development, evaluation, and deployment.

EMS must begin building the same—not as private versus public, paid versus volunteer, fire-based versus hospital-based, or rural versus urban—but as one profession.

The American Medical Association Model: Clear Roles and Shared Responsibility

The AMA's trustworthy AI framework[6] does not just call for "safe AI." It assigns responsibility across the ecosystem:

- **Developers** must build AI tools that are clinically validated, transparent, and explainable. They must ensure diverse, representative training data.

Most importantly, they must design tools that solve actual clinical problems—not engineer solutions in search of a problem.

- **Healthcare organizations** are charged with ethical deployment. Their role includes vetting vendors, integrating AI into clinical workflows, monitoring real-world performance, and supporting the clinical workforce through education and policy.
- **Physicians** remain accountable for final clinical decisions. AI can assist—*but never replace*—the clinician's judgment. AMA reminds us that licensed providers must evaluate whether any AI system
    - works as intended,
    - works for their specific patient population, and
    - improves outcomes in a measurable, evidence-backed way.

This is a mature, operational framework. Not just rules, but roles.

Translating This to Emergency Medical Services

Every AI-enabled system entering EMS—from documentation to dispatch to diagnostics—must answer these three questions:

- **Does it work?** Has the AI tool been validated for safety, bias, and accuracy under field conditions? Has it been tested with EMS protocols, not just hospital workflows?
- **Does it work for my patients?** EMS serves a unique mix of rural and urban patients who are multilingual, high-acuity, and socially complex. Has the tool been proven in that environment, or was it trained in a sterile academic setting?
- **Does it improve outcomes?** Does the AI help reduce scene times? Improve diagnostic accuracy? Enhance narrative quality? Or is it just an expensive distraction?

These questions should guide every EMS leader, agency, and medical director evaluating AI technology. If you do not ask them, someone else will. And the answers may not reflect your system, patients, or standards.

Building Our Guardrails

Here are a few initial steps that EMS must take now:

- **Adopt a shared framework of responsibility**: Developers, EMS systems, and clinicians each have a role. One cannot function ethically or effectively without the others.

- **Establish national criteria:** Just as the AMA has done, EMS must identify performance, equity, and operational metrics by which AI tools are judged.
- **Reinforce clinician authority:** AI should never override field judgment. Medical directors must ensure protocols preserve decision rights for licensed EMS professionals—even when AI disagrees.
- **Integrate AI education for the EMS workforce:** If clinicians do not understand what a model is doing, they cannot supervise it. EMS education must now include AI literacy.

The bottom line: The most dangerous AI systems are not systems that hallucinate or malfunction; they are the ones that function effectively but do not align with our needs.

Microsoft and Epic: What Emergency Medical Services Can Learn from Health Care's Artificial Intelligence Giants

While EMS has debated the "if," the rest of health care has been accelerating the "how."

Nowhere is this more visible than Microsoft's investment in AI, and they are leveraging that investment through a strong partnership with Epic Systems, the largest electronic health record (EHR) vendor in the United States. Together, they are not just integrating AI into the clinical experience—they are redefining it.[7]

Microsoft: Cloud Power Meets Clinical Intelligence

Microsoft entered healthcare AI with a long game in mind. Microsoft provides the computing backbone for scalable, secure, HIPAA–compliant AI infrastructure through its Azure cloud platform (fig. 6–2). But it's not just about servers. Microsoft's strategy centers on the following:

- **Natural language processing:** Allowing clinical systems to extract structured meaning from unstructured notes
- **Predictive models:** Assisting with patient risk stratification, sepsis alerts, and deterioration monitoring

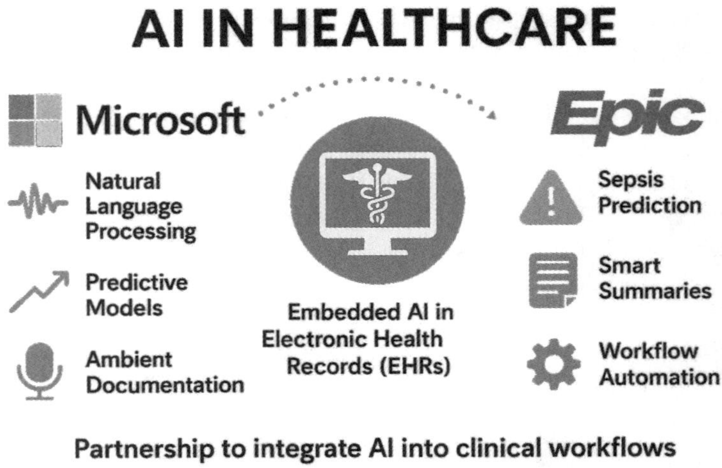

FIGURE 6–2. Major technology and healthcare companies are embedding AI directly into clinical workflows. Microsoft and Epic's partnership integrates natural language processing, predictive modeling, and ambient documentation into electronic health records, improving efficiency and diagnostic accuracy.

- **Ambient documentation tools:** Turning clinician–patient conversations into draft clinical notes with minimal interaction—already reducing burnout in primary care

Since 2015, Microsoft has poured billions into AI acquisition, research, and partnerships across the healthcare spectrum.[8]

Epic Systems: Embedding Artificial Intelligence Across the Care Continuum

Epic, managing health records for over 250 million patients, has integrated AI across multiple facets of its platform:

- **Sepsis predictive alerts:** Real-time risk notifications based on EHR data
- **Smart summaries:** Automatically generated patient handoffs using natural language processing and clinical logic
- **Workflow automation:** Identifying outliers, predicting length of stay, and optimizing orders

In 2023, Epic began deploying generative AI tools integrated with Microsoft Azure OpenAI Services, enabling providers to summarize visits, suggest follow-up plans, and draft documentation within the EHR interface.

Expanding Horizons: Agentic Artificial Intelligence and Beyond

At the Healthcare Information and Management Systems Society's Global Health Conference in 2025, Epic unveiled its investment in agentic AI, autonomous agents capable of performing tasks to enhance patient and clinician experiences. Epic has continued to expand its agentic AI footprint in every major product release since. These agents can[9]

- engage patients before appointments to gather goals and necessary information,
- identify and schedule pending tests before visits, and
- summarize information for both patients and clinicians, drawing insights from enterprise-wide data.

Epic has also integrated generative AI capabilities into its EHR, with approximately 125 features in development, including the following:

- **MyChart in-basket augmented responses:** Tool for drafting responses to patient messages, currently used by 150 healthcare systems and generates over 1 million drafts monthly
- **Clinician summaries:** Providing summaries for physicians and nurses across various use cases, from primary care to specialty-specific summaries and inpatient rounding

An Open Architecture That Fuels Innovation

Epic's evolving embrace of agentic AI is not just about smarter documentation—it's about strategic architecture. Epic has moved toward a more open ecosystem in recent years, aligning with the broader healthcare industry's push for interoperability and rapid innovation.

Epic has embraced an open ecosystem. Their growing developer showroom includes over 1,000 third-party applications integrated via application programming interfaces (APIs), with over 10 billion API calls monthly. This mirrors the same model that allowed Apple's App Store to revolutionize mobile innovation; Epic maintains the secure, foundational core but enables broad, rapid innovation through partnerships, and that is precisely what EMS lacks.

This shift mirrors the philosophy that made Apple's App Store a transformative force, acting not as the exclusive innovator but as the enabler of innovation. Epic remains the foundational engine for clinical documentation and billing. Still, it invites partners to build new tools on top of that platform, accelerating progress far beyond what any single company could deliver alone.

> *We hear a few different types of outcomes. Some of it is hard ROI [return on investment] ... Some is more of the soft ROI, like the feeling of reduced cognitive load or reduced burden... We hear all the time, physicians will say, "I could never go back to how things were done before."..."It saved my marriage," because they were no longer bringing charting home... I think we'll see a good set of outcomes around efficiency and time saved, and a good set of outcomes around workforce retention and job satisfaction across the clinical community.*
>
> — Seth Howard, Vice President of R&D at Epic[10]

That is not just a software upgrade. That is a cultural shift.

A Critical Lesson for Emergency Medical Services

AI does not wait for readiness. It scales where it's welcome.

EMS must confront a hard truth: We have historically maintained some of the highest barriers to entry in the healthcare technology landscape. The documentation tools, billing systems, CAD integrations, and electronic patient care reporting platforms we rely on are often proprietary, closed, and siloed. That may protect control, but it stifles innovation.

Epic's model shows a better way. EMS must move toward an open architecture that allows responsible innovation to plug in, experiment, and scale—without waiting on years-long contract cycles or a single vendor's roadmap. Without that shift, we will not just fall behind—we'll lock ourselves out of the tools that could save our workforce from burnout and help modernize prehospital care.

Multidisciplinary Alignment: American Nursing Association and American College of Physicians in Detail

While many national medical organizations have charted a responsible path forward for AI, the efforts of two organizations—the ANA and the ACP—have direct applicability for EMS. Each approach is tailored to the distinct needs of its workforce. Yet, it shares the same foundational values: safety, transparency, trust, and equity (fig. 6–3).

The Nursing Profession: Grounding Artificial Intelligence in Ethics and Human Dignity

In 2022, the ANA released its position statement, "The Ethical Use of Artificial Intelligence in Nursing Practice."[11] The position statement was neither vague nor optional—it was a clear framework for how AI must function in support of nurses and patients.

Key components include the following:

- **Patient privacy and confidentiality:** AI must never compromise HIPAA protections or ethical obligations to protect personal health information.
- **Informed consent:** Nurses are trained to ensure patients understand their care. AI systems must be integrated in a way that respects autonomy and informed choice.
- **Health equity:** AI should narrow disparities, not widen them. ANA emphasized the need to audit systems for bias and ensure they serve all populations.
- **Education and training:** Nurses must be prepared to not only use AI but also evaluate, challenge, and override it when necessary.

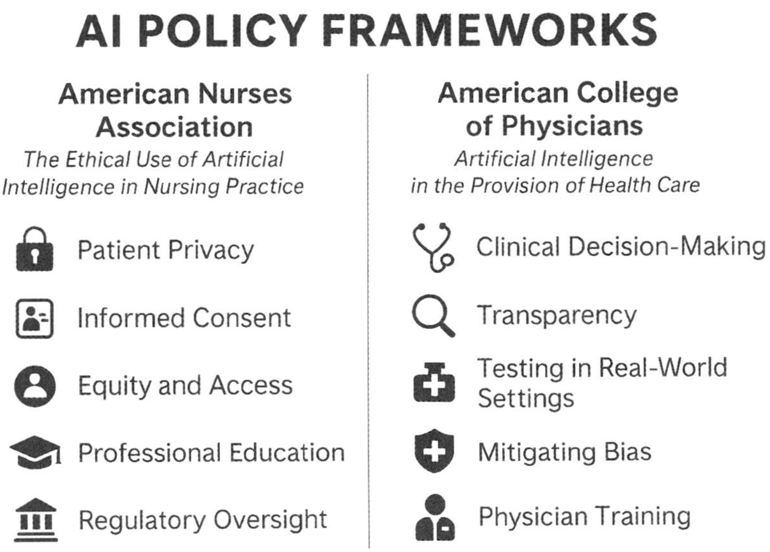

FIGURE 6–3. Professional associations are establishing ethical frameworks for the use of AI. The ANA and the ACP emphasize transparency, education, bias mitigation, and patient protection as essential to the responsible integration of AI.

ANA's position paper called for professional empowerment. According to the ANA, nurses must fully partner in designing, deploying, and governing the AI systems that will shape their practice.

The American College of Physicians Ten-Point Policy: From Concept to Compliance

In 2024, the ACP published one of the most comprehensive AI policies: "Artificial Intelligence in the Provision of Health Care."[12] This policy emphasizes a crucial point: AI is not neutral—it requires careful guidance.

ACP's policy outlines six key principles:

- **Real-world validation:** AI tools must be rigorously tested in diverse patient populations rather than relying solely on simulated models.
- **Anti-bias infrastructure:** Collaboration among developers, regulators, and healthcare systems is essential to effectively monitor and mitigate algorithmic bias.
- **Data privacy & security:** Safeguarding clinician and patient data against misuse and breaches is paramount.
- **Continuous feedback loops:** AI tools should be evaluated continuously to assess their clinical performance, equity, and unintended consequences.
- **Clinician education:** AI competency must be integrated into medical education and continuing education programs to prepare healthcare professionals for the future.
- **Governance:** Robust systems must be established to report and address AI-driven adverse events, similar to the protocols for pharmaceuticals and medical devices.

The strength of the ACP framework lies in its operational focus. It assigns clear responsibilities, establishes necessary guardrails, and prioritizes the clinician's role over the algorithm.

Shared Values, Shared Blueprint

Despite representing different clinical roles, the ANA and the ACP converge on four essential nonnegotiable values:

- **Transparency:** Clinicians and patients must know when and how AI technologies are utilized.
- **Safety:** AI tools must undergo rigorous validation, monitoring, and governance to meet safety standards.

- **Equity:** AI systems must be thoroughly evaluated for bias and designed to serve diverse populations fairly and effectively.
- **Trust:** AI must build confidence among clinicians rather than seek to replace their expertise.

These shared values align seamlessly with the public safety mission of EMS. They represent principles that EMS can—and must—embrace immediately.

What Emergency Medical Services Must Do Now

Vision is not enough. We have reviewed the frameworks and seen the national momentum. Now comes the question that matters most: What must EMS do—*right now*—to avoid being left behind?

If EMS does not define its AI future, others will do it for us, and we have seen this story before.

When Uber entered the transportation sector, New York taxi companies still held million-dollar medallions. They had the infrastructure, regulation, and legacy, but failed to adapt. They clung to control instead of designing the future. Within five years, the industry collapsed, losing two-thirds of its market share, not because service disappeared but because a model emerged that was more convenient, more transparent, and better aligned with the public's expectations.

What if the same happens to EMS?

Yes, there will always be a need for medical response and transport—but that future might operate through a model we do not control or even recognize. It may be redefined by private logistics companies, AI dispatch platforms, or autonomous transport networks that meet public expectations in faster, cheaper, or more flexible ways.

> *EMS agencies still have a choice: lead, embrace, and design, or wait to be disrupted. The agencies that lead will define what healthcare-integrated EMS looks like. The agencies that wait will operate within the results, and those results may not align with their workforce, traditions, workflows, ethics, or mission.*

Want the Full Emergency Medical Services Artificial Intelligence Framework?

For detailed implementation guidance—including ethical principles, clinician authority, national interoperability, and data integrity safeguards—see chapter 11.

That chapter presents the EMS Conceptual Framework for Ethical and Responsible AI, featuring 12 field-tested pillars, operational safeguards, and the national alignment EMS must demand.

Chapter 6 tells us why EMS must lead. Chapter 11 shows how.

Key Terms and Definitions

- **Augmented Intelligence:** A term the AMA promoted to emphasize AI systems that enhance, rather than replace, clinician decision-making.
- **Agentic AI:** AI systems that can autonomously perform tasks and initiate actions based on learned objectives, potentially transforming administrative and clinical workflows.
- **Ambient Documentation:** Using AI to transcribe and structure clinician-patient conversations in real time reduces the burden of manual charting.
- **Ethical AI:** AI tools designed and governed to uphold principles of transparency, equity, data privacy, and clinician oversight in healthcare settings.
- **Open Architecture:** A system design that allows third-party software or tools to integrate via standardized APIs, enabling innovation across platforms, which is critical for scalable EMS technology advancement.
- **Unified Framework:** A nationally coordinated policy or operational standard adopted by a profession (e.g., medicine, nursing) to guide ethical, clinical, and technical integration of AI.
- **Silent Encroachment:** The phenomenon where external actors (e.g., tech companies or investors) build AI solutions for EMS without input from the EMS profession, leading to tools that may not reflect EMS needs, values, or regulatory structures.
- **AI Literacy:** Foundational understanding of how AI systems work, their risks, and appropriate clinical oversight—now considered essential for clinician education and professional competence.

Reflection Questions

1. What is the risk of continuing without a unified EMS position on AI?
2. Who within your organization or association can lead or contribute to national-level guidance?
3. How might AI help reduce workload, burnout, or inefficiencies in your current EMS environment?
4. What role should state EMS offices, educators, or regulators play in shaping AI policy and training?
5. What values or safeguards might be missing if a private company built tomorrow's "next-generation EMS platform" without EMS input?

Study Questions

1. According to the chapter, what is the main reason EMS has been left out of national AI strategy conversations?
2. Which of the following best describes the federal government's current stance on AI under both the Biden and Trump administrations?

   a. AI is primarily a concern for defense and national security, not health care.
   b. AI is viewed as a national strategic asset, though policy emphasis varies by administration.
   c. AI adoption is being paused until further ethical reviews are complete.
   d. AI development has been fully delegated to private industry without federal oversight.

3. Which of the following best captures the AMA's position on the use of AI in health care?

   a. AI should replace repetitive clinical tasks without physician oversight.
   b. AI must reduce cost above all other priorities.
   c. AI must enhance, not supplant, clinical decision-making and align with medical ethics.
   d. AI tools should be deployed before being validated to ensure innovation speed.

4. What term did the AMA adopt in 2019 to emphasize AI's role in supporting—rather than replacing—clinicians?

5. What three critical questions should EMS leaders ask before adopting any AI-enabled system?
6. List two essential principles shared by both the ANA and ACP regarding responsible AI in health care.
7. What is "silent encroachment," and why is it a risk for EMS in the age of AI?
8. Describe how the partnership between Microsoft and Epic Systems has changed the way AI is integrated into clinical workflows.
9. What does the chapter recommend as the most urgent leadership action EMS must take to ensure inclusion in national AI policy?
10. What lesson can EMS learn from the taxi industry's failure to adapt to technology platforms like Uber?

CHAPTER SUMMARY

Engage or Be Erased

AI is no longer speculative—it is funded, operational, and embedded in the national infrastructure of modern health care. The AMA, ANA, ACP, and AAFP have all taken decisive steps to define how AI must be integrated, with ethics, equity, and clinician empowerment at the core. They have built standards, published roadmaps, and shaped national policy; EMS has not.

As federal agencies release AI executive orders and technology giants like Epic and Microsoft redefine clinical workflows, EMS remains absent from the discussion. In this era, silence is not neutral—it is exclusion. If EMS does not define how AI fits into our profession, the rules, tools, and expectations will be defined without us.

This chapter does more than sound the alarm. It sets the stage for the Conceptual Framework for Ethical and Responsible AI in EMS introduced in chapter 11—a twelve-pillar roadmap grounded in real-world practice and the nonnegotiable truths that define EMS.

The path forward is not technical—it is leadership-driven.

If EMS wants to shape the next era of emergency care, we must

- Establish a unified national framework for AI in EMS—with shared governance across our profession,
- Integrate AI ethics and literacy into EMS education, certification, and continuing professional development,
- Ensure real-time clinical interoperability, connecting EMS to hospital EHRs, health information exchanges, and system-wide analytics,
- Demand open architecture and vendor transparency, so EMS professionals can design, test, and lead innovations that reflect our operational realities, and
- Engage in national AI policy discussions, ensuring EMS voices are included where the future is being written.

AI is not a tool we can ignore. It is the operating environment of health care's future. And unless EMS defines its standards, values, and priorities now, we will find ourselves practicing in a system built without us.

Presence matters. Leadership matters. This is our moment to act.

7

Artificial Intelligence–Powered Documentation in Emergency Medical Services

LEARNING OBJECTIVES

After reading this chapter, the reader will be able to

- Explain the legal, operational, and clinical importance of the EMS patient care narrative, especially its role in administrative and regulatory investigations.
- Trace the historical development of EMS documentation, from Dr. Safar's 1969 vision to the implementation of NEMSIS and the current role of structured data in public health and national security.
- Differentiate between structured data and narrative documentation, and articulate the unique value each provides in clinical, legal, and operational contexts.
- Describe the capabilities and emerging role of large LLMs in EMS documentation, including voice-to-narrative tools, real-time clinical prompts, and QA enhancements.
- Evaluate common barriers to AI documentation adoption in EMS—technical debt, cultural resistance, lack of standards, and underfunding—and develop strategies to overcome them.
- Articulate the clinician's responsibility in AI-powered documentation, including legal authorship, professional accountability, and narrative verification.
- Assess how AI documentation tools can reduce burnout, increase efficiency, and enhance clinical decision-making, while preserving the authenticity of the EMS clinician's voice.
- Discuss the strategic importance of integrating EMS documentation into the broader healthcare data ecosystem and how AI can serve as a bridge to improve interoperability, patient outcomes, and professional visibility.

Prologue: The Power of the Narrative

In the courtroom, at the complaint desk, or before a licensing board, the story rarely starts with sirens: It begins with the narrative.

As a former state EMS director in two states, a 30-year paramedic, and a regulator for over 15 years, I can confidently say that few things are more powerful (or more perilous) than your patient care report—not your equipment, not your words, not even the protocol you followed.

When a complaint reaches a regulator's desk, the review begins with one of our most immediate legal tools: the EMS narrative. If that documentation is clear, complete, and professional—if it shows the scene and explains the decision-making—it can stop an investigation before it starts. But if it is sloppy, vague, or missing key context, the EMS clinician begins that process at a disadvantage. Fair or not, that is the reality.

What makes the EMS narrative uniquely powerful is its timing. It is contemporaneous evidence—written at the time of care, under real-world conditions, and long before any complaint, lawsuit, or review is ever filed. It is created without hindsight or legal preparation as a clinical record, making it profoundly influential.

Unfortunately, most legal sections in EMS textbooks still emphasize abandonment, negligence, or criminal action—rare events in modern practice. The most common source of legal and regulatory exposure for EMS clinicians is not a courtroom drama. It is an administrative review. And the determining factor in many of those cases is documentation quality.

Fast forward to 2023. I helped develop the first EMS-specific large language model (LLM) for AI-powered documentation in partnership with Wavenet, a Singapore-based technology firm. Our goal was to create faster reports and build more accurate, complete, and clinician-friendly documentation. The technology worked, but the profession was not ready. Resistance was high, buy in was slow, and the tools were ahead of the culture.

By 2025, that changed. Major electronic patient care reporting (ePCR) companies began integrating AI-powered documentation tools. The concept we had championed finally took root, but did not grow from within. EMS had the opportunity to lead. Instead, we sparked the idea and watched others scale it.

That experience carries a lesson: Documentation is not just about data entry; it is about the narrative. And AI must be trained to understand that.

EMS is one of the only medical professions that operates in uncontrolled, unsterile, unpredictable environments—hoarder homes, icy ditches, freeway shoulders, barking dogs, screaming families. If AI-generated narratives fail to capture that complexity, they fail, no matter how perfect the grammar may be.

Contemporaneous Evidence in EMS Reports

Contemporaneous evidence refers to documentation created at the time of an event, rather than after-the-fact. In EMS, this means the patient care narrative is authored during or immediately after the incident—often while events are still fresh in the clinician's mind and before a complaint was filed.

Good documentation enhances

Credibility: Created before any complaint, lawsuit, or investigation, the report carries weight as an unbiased and factual record.

Legal Protection: Courts and regulatory agencies give significant weight to contemporaneous documentation, as it reflects real-time decision-making.

Professional Defense: A complete, accurate narrative can protect EMS clinicians from unfounded allegations by clearly outlining observations, decisions, and actions taken.

A strong EMS narrative is not just a chart—it's your first line of defense. Write it like it will be read in court, because one day it might be.

This chapter is not just about innovation. It is about preservation. We must protect the clinician's voice. We must ensure that AI reduces the documentation burden—but never replaces the human truth of what happened at that moment, at that scene, on the front lines of care.

When the complaint comes in, the narrative is often all we have.

Dr. Safar's Legacy and the Birth of Emergency Medical Services Data

In 1969, Dr. Peter Safar stood before the Airlie House Conference and proposed a radical, standardized, computerized system for documenting every ambulance call. At the time, it was a visionary concept. There was no national or standard curriculum for emergency medical technicians (EMTs), no national registry, and minimum equipment list for ambulances. The infrastructure did not exist. But Safar saw where EMS could go, and saw its potential to turn into a data-driven, accountable, and integrated branch of modern medicine.

Safar was more than a clinician. He was a futurist, an architect of systems, and a relentless innovator. He helped invent critical care medicine and modern resuscitation. He coauthored the EMT Paramedic curriculum with Dr. Nancy Caroline. He founded the Freedom House Ambulance Service, the first program to train lay Black citizens to deliver care at a standard equal to—or in some cases exceeding—that of physicians at the time. And amid all that, he still looked further toward *documentation*.

When I began my EMS career in the early 1990s, my first paper run sheet was nearly identical to what Safar had proposed decades earlier. That alone speaks volumes about how far ahead he was.

Today, his vision is not only realized—it is scaled. More than 60 million[1] EMS patient care reports are submitted annually to the National EMS Information System (NEMSIS), forming the largest near-real-time prehospital dataset in American health care. That data now powers public health dashboards, overdose surveillance, hospital surge readiness, disaster response, syndromic detection, and national revisions to EMS Scope of Practice. It is embedded into federal systems and increasingly used to inform policy.

Yet for many clinicians in the field, NEMSIS still feels like paperwork, not progress. That disconnect is not due to a lack of value. We have built systems that still place the full documentation burden on the individual clinician.

Safar anticipated this, too. He understood that EMS would need to document not only vitals and timelines but the clinical and situational context. He knew we would eventually need help doing it. That day is no longer on the horizon. That day is now.

Safar was not just a clinician but a futurist, visionary, and systems architect.

Emergency Medical Services History Spotlight: The Airlie House Conference—Where Emergency Medical Services Was Born

In 1969, a landmark meeting was convened at Virginia's Airlie House Conference Center. It brought together physicians, military leaders, federal officials, and early EMS pioneers to answer one pressing question: What should emergency medical care look like in the modern era?

The result was a blueprint that still defines the foundation of EMS today.

Among the most important outcomes of the Airlie House Conference are as follows:

- **Standardized patient care documentation:** Dr. Peter Safar's proposal for a unified ambulance run report laid the groundwork for what would eventually become NEMSIS.
- **Creation of the National Registry:** The group endorsed the idea of national certification for EMS clinicians, resulting in the formation of the National Registry of EMTs in 1970—a turning point for consistency, quality, and recognition across the country.
- **Modern ambulance design:** Participants rejected the outdated use of converted hearses. They called for purpose-built ambulances with space, equipment, and safety features appropriate for delivering care, marking the beginning of today's national ambulance standards.
- **National curriculum foundations:** The conference initiated the effort to create standardized EMS education and certification. It set the stage for consistent training, clinical protocols, and clear professional pathways.
- **Federal recognition of EMS as a system:** Most critically, the Airlie House Conference reframed EMS as more than transportation. It affirmed EMS as a system of care, deserving of national coordination, investment, and policy leadership.

The impact of Airlie House cannot be overstated. Nearly every central pillar of EMS today—standard documentation, national certification, ambulance design, clinical protocols, and system integration—can be traced back to that conference. It was a rare moment of clarity, foresight, and national alignment that continues to shape the profession's trajectory.

Structured Data, Unmatched Potential

What Dr. Safar envisioned in 1969 is no longer aspirational—it is operational nationally. NEMSIS, developed by the National Highway Traffic Safety Administration, is now the technical backbone of EMS data collection and analysis across the United States.

More than 60 million EMS activations are now submitted annually to NEMSIS, capturing clinical care from diverse scenes—cardiac arrests, vehicle collisions, overdoses, and beyond. This makes it the most extensive, fastest, and comprehensive prehospital dataset in American health care. But NEMSIS is more than a repository—it is a transformation tool.[2]

CASE STUDY

2003—When Emergency Medical Services Surveillance Met National Security

In the wake of the September 11th attacks, the U.S. Department of Homeland Security launched the Urban Areas Security Initiative (UASI), designed to enhance emergency preparedness in high-risk metropolitan regions. Most EMS agencies nationwide were still operating on paper documentation at that time. However, a bold idea shaped Indianapolis and Hamilton County, Indiana.

As a suburban EMS chief during that era, I was part of a multidisciplinary team that asked a radical question: *What if EMS data could help protect the country?* Our concept was simple but groundbreaking: digitize EMS patient care reports, link systems across jurisdictions, and share data across EMS, hospitals, and public health agencies in near real time. Then, the system can be used to define what "normal" looks like, so any emerging abnormal pattern can be detected immediately.

With support from UASI funds, we built one of the first large city interoperable EMS surveillance networks in the United States (fig. 7–1). It operated before smartphones were common, national dashboards existed, and AI had entered the conversation. Yet the logic was sound: If respiratory complaints spiked unexpectedly or clusters of seizures or overdoses appeared across zip codes, the system could recognize those anomalies and send immediate alerts to decision-makers via text or email.

This was not a theoretical exercise. The system worked. It flagged outliers, initiated coordinated responses, and most importantly, shifted the perception of EMS from a transport function to a strategic public health and homeland security asset.

We did not just create better data—we reframed the role of EMS in protecting the nation.

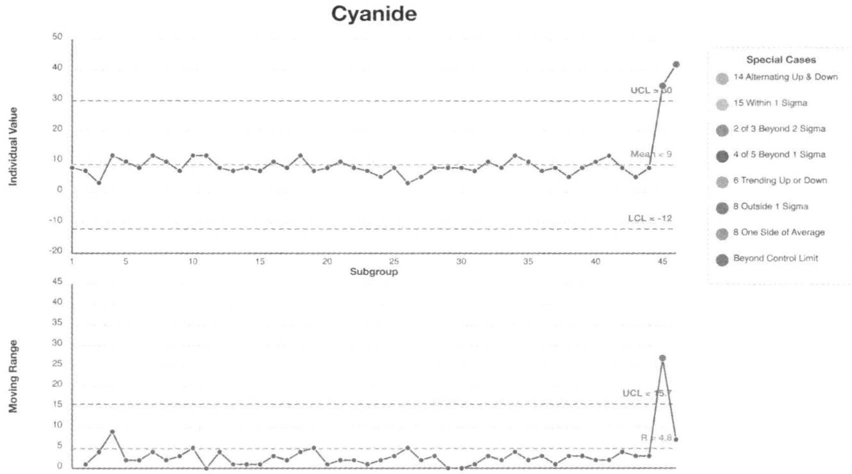

FIGURE 7–1. This chart represents an early conceptual model of syndromic surveillance developed during the 2005 Indianapolis–Hamilton County UASI ePCR project. It marked the first large-scale deployment of ePCR in EMS, demonstrating how EMS data could be leveraged for national security monitoring and the early detection of emerging public health threats.

Structured ePCR fields enable powerful capabilities:

- **Public health monitoring:** Real-time tracking of opioid overdoses, heat-related illness, and emerging disease trends.
- **Syndromic surveillance:** Early alerts for respiratory clusters, chemical exposures, or bioterrorism events.
- **National security preparedness:** Anomaly detection that can identify patterns signaling mass casualty incidents or coordinated attacks.
- **Clinical and operational standards:** Data-driven revisions to Scope of Practice models, protocol refinement, and benchmarking for system improvement.

No medical profession produces data with this scale, speed, and operational detail. NEMSIS enables everything from local quality assurance (QA) to federal situational awareness. Yet despite these capabilities, clinicians often see only the burden—endless fields, dropdowns, and form logic that must be completed after intense, chaotic calls.

The system's power is undeniable, but it comes at a cost.

The Cost of This Value

The power of EMS data is undeniable. It drives national surveillance systems, informs policy, supports hospital readiness, and enables AI to operate in real time. But that power comes with a cost—a burden carried almost entirely by EMS clinicians in the field.

Each EMS report includes hundreds of structured fields: timestamps, vital signs, treatment codes, incident narratives, patient assessments, and more. Every field improves visibility for the system. But every field adds stress for the clinician.

Unlike hospital environments, EMS clinicians usually start documentation while in motion—often under pressure, in poor lighting, with limited space and no administrative support. A paramedic may finish resuscitation on the side of a highway, then crouch in a freezing ambulance to input times, medications, assessments, and summaries with a headlamp or backlit keyboard before responding to the next call.

Even as user interfaces have improved, the reality remains: EMS documentation is among the most cognitively demanding, time-intensive tasks clinicians face—and it happens at the end of their most exhausting moments.

Structured documentation is essential to national strategy. However, the price has been burnout, frustration, and, in many cases, reduced clarity in the narratives that should protect the clinician.

Suppose EMS documentation powers public health dashboards, clinical research, quality benchmarks, disaster detection, and AI integration. In that case, the burden of that infrastructure should not rest solely on the backs of frontline responders.

This is not just a workflow issue. It is a workforce sustainability issue.

Artificial Intelligence in Clinical Documentation: A Healthcare Standard, an Emergency Medical Services Gap

Nowhere is AI's impact more visible—or more urgent—than in clinical documentation. What once consumed hours of typing and postshift mental fatigue is now being reimagined through ambient AI scribes and LLMs trained on real clinical workflows.

These systems "listen" to the clinical encounter, structure the documentation in real time, and generate a draft summary before the patient leaves the room. The result? Clinicians spend more time in conversation and less time chasing charts.

Nearly 30% of physicians in nonacademic settings use ambient documentation systems.[3] That number is skyrocketing, driven by a single reality: The documentation burden is no longer sustainable.

Physician burnout—already a crisis in primary care, emergency medicine, and behavioral health—is directly tied to time spent documenting after hours. AI is now being used not only as a clinical tool but also as a workforce wellness intervention. And it works.

In early studies,

- documentation time dropped by 50–70%,[4]
- clinician satisfaction scores improved, and
- note completeness and standardization increased.[5]

These tools are not experimental. They are billing-compliant, HIPAA-secure, and rapidly becoming standard in major healthcare networks. And yet in EMS, documentation remains frozen in time. While hospitals deploy ambient scribes, most EMS clinicians still rely on

- manual charting systems with inconsistent structure,
- lengthy QA loops that delay final reports, and
- frustrated clinicians typing narratives at the end of the shift, often from the ambulance bay.

The disconnect is no longer about technology. It is about culture, courage, and what we are willing to invest in. AI is already reducing liability, improving narrative clarity, and restoring time to clinicians in other settings. EMS cannot afford to ignore this momentum.

Documentation is not just a clerical task—it is legal protection, clinical continuity, and operational visibility. The longer we delay this transformation, the more we widen the gap between EMS and the healthcare system we claim to be part of.

From Theory to Reality

In 2023, the White House launched the National Dashboard for Nonfatal Overdoses, built on real-time NEMSIS data (fig. 7–2).[6] This was more than a public health milestone—it was a turning point for EMS visibility. It confirmed that EMS data is not just passively collected; it actively drives national decision-making.

EMS documentation has been considered operational paperwork for decades, used locally, and rarely considered in broader healthcare strategies. That

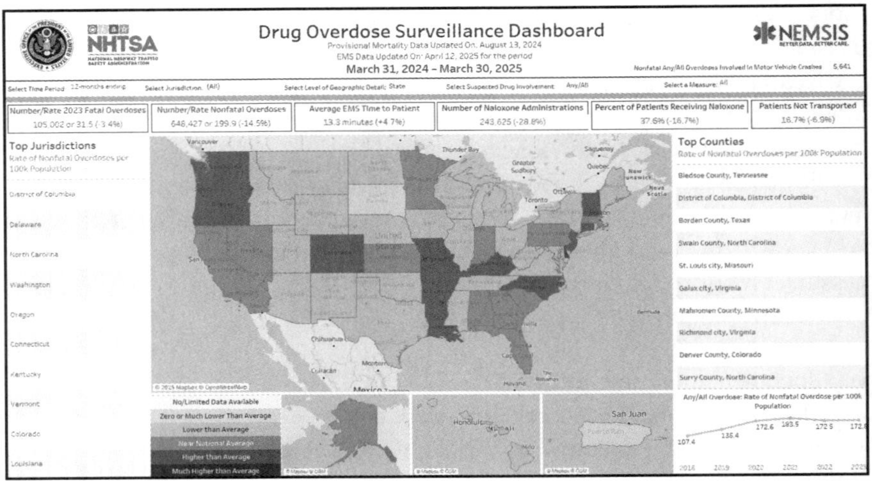

FIGURE 7–2. National EMS data now drives federal overdose surveillance. Real-time analytics from NEMSIS enable the tracking of nonfatal overdoses, naloxone use, and geographic trends, demonstrating the public health value of EMS data integration.

narrative has shifted. Today, structured EMS data supports overdose surveillance, pandemic response, syndromic detection, and disaster coordination. It informs policy at the federal level. It shapes hospital preparedness. It also validates EMS's essential role in the national health and security infrastructure.

However, this progress in structured data tells only half the story.

The next frontier is the patient care narrative—the part of the record explaining the chaos, judgment, and nuance of clinicians' experiences. As AI continues to evolve, the systems that succeed will not be those that collect the most data. They will be the ones who interpret the data in its true context.

This is where AI, especially LLMs, enters the equation—not just to summarize what happened, but to preserve why it mattered. In the following sections, we move from policy to practice—from the national shift to AI-powered documentation to the local realities of how EMS must respond.

The Missing Story: Why Narratives Still Matter

Structured data (all those check boxes and dropdown boxes) gives systems what they need to track events, calculate trends, and populate dashboards. But structured data alone cannot explain what actually happened, under what conditions, or why certain decisions were made. That context—the human reality of the call—lives in the narrative.

A dropdown field might show that fentanyl was administered, but only a well-written narrative can explain that it happened in a collapsed hallway next to a crying child, while the patient lay partially trapped beneath debris. It will document that the EMT crawled through wreckage, cut through a sweatshirt, and improvised an IV line while navigating broken furniture. It will explain that a 19-minute delay in obtaining a blood pressure was due not to negligence, but to a combative patient, a hoarder's home, a bystander yelling on a narrow mountain road in the rain, and a medic slipping on a soaked staircase while carrying a Lifepak and managing hypoglycemia.

Structured data records actions—narratives document judgment.

The EMS narrative is not a secondary detail. It is central to clinical accountability, regulatory review, and legal defense. It reflects the scene, the decision-making process, and the professionalism of the clinician under pressure. It tells the story of care in real time—unfiltered, undocumented by others, and created long before anyone asks questions about what occurred.

As a former state EMS director and regulator, I have seen written narratives clarify complex events, confirm competent care, and close investigations without further action. I have also seen vague, rushed, or poorly written narratives spark unnecessary disciplinary reviews—even when the care was clinically appropriate. The quality of the narrative directly influences the outcome of that review. It speaks when the clinician cannot.

When the record reads with clarity, coherence, and professionalism, it builds confidence. When it reads like an afterthought, it undermines it. Whether fair or not, that is the standard under which EMS clinicians are judged. These narratives are not just documentation—they are the only enduring voice of the EMS clinician once the call is over.

What Narratives Do That Data Can Not

Structured data supports interoperability. It enables billing, fuels research, powers public health alerts, and provides the quantitative backbone of healthcare surveillance. But it does not explain clinical decisions, document uncertainty, ethical dilemmas, or environmental complexity, or justify why an IV was deferred, transport was delayed, or a procedure was attempted twice. Narratives do.

A well-written EMS narrative captures the clinician's thought process. It describes the scene: the patient who refuses care despite hypoxia, the apartment door that took five minutes to breach, the confusion caused by conflicting bystander accounts. It provides a complete picture that legal counsel, hospital staff, QA reviewers, and regulators rely on when evaluating care.

Narratives serve five critical functions. They provide context for clinical decisions. They document obstacles such as environmental hazards, hostile scenes, or patient behavior. They reflect the rationale behind split-second judgments. They reinforce professional tone, conveying confidence, compassion, and clinical presence. And most importantly, they serve as contemporaneous legal evidence—a record of what happened, created in the moment, before hindsight or legal interpretation reshapes the facts.

AI can support this process. It can suggest structure, prompt for clarity, and help clinicians capture details they might otherwise forget. But AI cannot witness the scene itself. It cannot replace the clinician's lived experience. The power of the narrative comes from its authenticity—from a human describing what another human endured.

The stories of the EMTs and paramedics matter because once the call ends, the story is all that remains.

The Rise of Artificial Intelligence Documentation

In 2023, while most of the EMS profession remained uncertain—or even unaware—of AI's real capabilities, one project moved quietly ahead. In partnership with Global Wavenet, a Singapore-based technology firm, we designed and deployed a prototype EMS documentation system grounded in real-world pressure, not academic theory.

This was the first known use of a LLM integrated into EMS workflows using Microsoft's HIPAA-compliant enterprise cloud. The system combined voice-to-text input, structured clinical logic, and generative AI to build authentic narratives in real time. Its purpose was not just speed. It was to restore the lost art of storytelling in EMS documentation—something long buried beneath dropdowns and form fields. And it worked.

Field clinicians could speak freely while treating patients, capturing their observations and decisions without distraction. The system translated that speech into clean, accurate narratives in real time. Clinicians could document for the first time in decades while staying fully present during patient care.

But despite early validation, adoption stalled. Budgets were tight. Vendors cited "technical debt"—outdated software not designed for natural language processing. Some vendors simply said, "No one will buy this. No one is asking for it."

The challenge was not technological. It was cultural. EMS leaders, agencies, and vendors lacked a shared vision for innovation. The tool did not fail—imagination did.

That early concept, however, planted a seed. By 2025, nearly every major ePCR vendor had announced plans or begun integrating LLM-based documentation tools. What once felt radical is becoming standard: voice-to-narrative engines, smart QA prompts, and real-time clinical decision support.

And let us not misremember history—this did not begin with Silicon Valley. It started inside EMS. EMS professionals envisioned it, built it, and tested it first. The rest of the story is still being written, and it belongs to EMS clinicians who choose to stop waiting for permission and start shaping the systems they work within, rather than remaining passengers in decisions that define our profession.

What a Narrative Does That No Dropdown Ever Will

Structured data records.

- BP is 136/88; pulse is 92; and respiratory rate is 12
- 12-lead acquired
- IV started
- Transported Code 2

The narrative should tell the why.

> "Patient was found lying supine in a cluttered hallway, alert but pale and diaphoretic. Vitals were delayed due to combative behavior. The family on the scene is hysterical. Advanced life support interventions initiated en route amid limited access and loud ambient noise."

Structured data powers analytics. Narratives protect clinicians.

When the complaint comes months later, only the narrative will explain the decisions, the delays, and the scene. It is your defense. It is your voice.

Features and Promises of Large Language Model-Based Documentation

AI-powered documentation is not just about convenience—it is about restoring clinical focus, reducing liability, and unlocking efficiency at scale. LLM-enabled tools represent the most transformative leap in EMS charting since the shift from paper to ePCR. This is not merely a technical upgrade but a reimagining of how we capture, organize, and verify prehospital care records in real time.

These tools are not designed to replace EMS clinicians. They are built to amplify their expertise and protect their time.

Core capabilities of modern LLM-based EMS documentation systems include the following:

- **Voice-to-field input:** Clinicians narrate while treating the patient. AI converts this into complete narratives, organizing it according to standard documentation formats like SOAP (structured subjective, objective, assessment, and plan format), Chartnote, or DRAATT (dispatch, response, arrival, assessment, treatment, and transport).
- **Multispeaker recognition:** The system distinguishes between EMS crew members and captures relevant dialogue, ensuring accuracy even in dynamic, multiprovider scenes.
- **Real-time clinical prompts:** The AI can ask questions or highlight missing elements as information is processed, such as, "You mentioned chest pain. Would you like to include the 12-lead electrocardiogram results?" This ensures that documentation supports both clinical best practices and legal defensibility.
- **Autocompletion of fields:** By analyzing the narrative, AI can populate related structured fields—like vitals, impressions, or interventions—saving valuable time and reducing documentation fatigue.
- **Seamless integration with ePCRs:** These tools are no longer standalone experiments. Leading vendors are embedding LLM capabilities directly into their platforms, enabling clinicians to generate, review, edit, and approve their reports in the same system they have used.
- **Accuracy and compliance:** Advanced medical spell-check, standardized terminology, and HIPAA-compliant encryption ensure that reports meet clinical and legal standards.
- **Efficiency and burnout reduction:** Early pilots report time savings of 10–20 minutes per call. Multiplied across a busy shift or system, this translates into more restocking, report handoff, rehab, and patient care time.

However, the promise does not end with time saved. These tools unlock new potential for the following:

- **Improved quality assurance/quality improvement (QA/QI):** Natural language processing allows QA teams to search for patterns, flag omissions, or validate protocol adherence more efficiently.
- **Billing accuracy:** Clean, legible, structured reports reduce reimbursement denials and improve revenue cycle efficiency.
- **Legal protection:** A clear, complete, and contemporaneous narrative remains your first line of defense. With LLM assistance, clinicians are no longer alone in crafting it under pressure.

Adopting AI is a strategic augmentation that restores what matters most—your time, voice, and professional integrity.

Barriers to Adoption and Why They're No Longer Excuses

When AI-powered documentation first appeared in EMS, the resistance was swift and predictable. Leaders cited limited budgets, technical limitations, and a lack of apparent demand. Vendors pointed to outdated system architecture. Clinicians understandably questioned whether these tools would work in the field, under pressure, and without compromising the care they deliver.

All of those concerns were real, but none of them remain valid:

- **Technical debt is temporary:** Many legacy ePCR platforms were built for checkbox logic, not natural language processing. That was a design limitation, not a permanent one. Today, leading vendors are rewriting their infrastructure to accommodate LLM integration. The question is no longer if this capability exists. It is determined by whether your current vendor is capable of delivering it. If not, it is not your outdated documentation but your platform.
- **Lack of standards is a leadership gap:** Unlike other healthcare sectors, EMS has not yet defined shared expectations for AI in documentation. There are no consensus-based frameworks, certification pathways, or evaluation metrics. But that is not a reason to delay. It is a reason to lead. Other professions built their standards before mandates arrived. EMS must do the same.

- **Budget constraints are a known reality—but not an excuse:** Yes, EMS is underfunded. But every hour spent on inefficient documentation is time lost to fatigue, burnout, delayed QA, or missed revenue. Investing in LLM-enabled tools is not a luxury—it is a workforce retention strategy and a pathway to quality.
- **Culture is the real challenge:** More than infrastructure or dollars, EMS struggles with a deep-rooted resistance to change. We are a conservative profession, and we often equate familiarity with safety. But outdated workflows create stress, reduce quality, and undermine retention. This is no longer a conversation about tools. It is a test of leadership.

Every barrier cited in 2023 is solvable in 2025. The technology is viable, the use cases are proven, and the need is urgent. The only remaining barrier is our willingness to act.

The question is not whether EMS can adopt AI-powered documentation.

The question is: Will we build the future or remain anchored in the assumptions of the past? We have the opportunity to lead with clarity, courage, and collaboration.

The Future of Documentation Is Not Just Faster—It's Smarter

Speed alone is no longer the benchmark of innovation. In the next era of EMS documentation, the accurate measure of progress is intelligence—how effectively a system supports clinical thinking, captures context, and preserves legal integrity. The shift is not from manual to automatic; it is from isolated charting to active partnership.

AI-powered documentation is evolving beyond transcription. The next generation of LLMs will function as scribes and real-time clinical collaborators. They will prompt, validate, and surface insights that help EMS clinicians deliver safer, more complete care and generate cleaner, more defensible records.

Imagine a system that listens to your narrative in real-time and responds with clinical prompts:

> "You mentioned chest pain, but no 12-lead electrocardiogram has been documented. Was one performed?"
>
> You reply, "No—the monitor failed," and the report immediately captures that clarification.

This is not hypothetical. These capabilities are already being deployed in hospital settings, and they are starting to be brought to EMS now.

LLM-assisted documentation will soon offer these:

- **Instant QA:** Flagging high-risk narratives (e.g., pediatric seizures, refusals, AMA calls) before submission.
- **Clinical decision support:** Offering real-time protocol-based reminders, tailored to working impressions and patient presentation.
- **Medication and interaction checks:** Reviewing drug choices against patient condition, flagging contraindications before errors occur.
- **Automated ePCR field completion:** Extracting structured elements like vitals, chief complaint, and treatment timelines from the narrative.
- **Integrated hospital notifications:** Generating a summary and ETA for transmission to the emergency department, streamlining handoffs, and reducing radio traffic.
- **Billing optimization:** Ensuring documentation includes transport justification, procedural detail, and other billing-critical data points.

What Vision Costs

Many EMS leaders tell me they do not have the money.

I hear it all the time at conferences—"Donnie, this all sounds great, but we're still holding bake sales just to afford diesel for our ambulances." And I get it. I've been there. I've held the bake sales. I know what it's like to choose between equipment and payroll.

But let me offer you a story we forget too easily. Decades ago, Dr. Peter Safar—one of the founding figures of EMS—envisioned a future where every EMS service would document patient care electronically. At the time, a single computer cost two to three million dollars in today's money. There were no ePCRs, no servers in the station—just a radical vision.

Safar didn't wait for technology or funding to catch up. He called his shot, and because of that, he gave EMS a direction to grow toward. Today, we have NEMSIS, cloud-based platforms, and AI tools writing reports in real time.

The truth is that vision costs nothing. But without it, EMS pays the highest price—irrelevance. The challenge is real, but that can never be an excuse to stop imagining. Because if we don't lead the future of EMS, someone else will.

These tools do not diminish the clinician's role—they expand it. They reduce the risk of omissions, ensure documentation integrity, and give time back to the field. When implemented correctly, AI transforms documentation from an administrative burden into a clinical asset.

This is the path forward—not just faster, smarter, safer, and more aligned with what clinicians need and the healthcare system demands.

The Win-Win-Win

Adopting AI-powered documentation is a technological advancement and a strategic improvement that benefits every stakeholder in the EMS ecosystem (fig. 7–3).

For the *EMS clinician*, it means less time typing and fewer missed details. Real-time prompts, voice-to-narrative tools, and automated field completion reduce cognitive load and preserve critical thinking capacity

WHAT HAPPENS WHEN WE WAIT?

EMS cannot afford to remain passive in the face of documentation innovation. The cost of inaction is not just inefficiency—it is professional invisibility.

Delaying AI adoption means

- Clinical innovation bypasses our systems.
- Field clinicians burn out under outdated workflows.
- Critical context is lost in templated reports.
- QA/QI lags behind healthcare standards.
- EMS remains disconnected from hospitals and payers.

If EMS does not lead, others will decide how we document—and what that documentation says about us.

FIGURE 7–3. Failing to modernize EMS documentation risks professional invisibility. Without proactive AI adoption, EMS falls behind healthcare standards, loses clinical context, and remains disconnected from hospitals and payers.

for patient care. Instead of spending the end of a shift reconstructing scenes, clinicians gain time for recovery, debriefing, or operational readiness. More importantly, they gain a safety net—a system that helps ensure nothing essential is overlooked.

For the *medical director*, this means more consistent, legible, and complete documentation. Innovative QA systems can flag outliers, suggest improvements, and track protocol adherence. This supports clinical governance, reduces liability, and allows for data-driven refinement of treatment pathways. AI does not replace oversight—it enhances it.

For the *EMS agency*, it means faster reporting, improved data quality, and greater operational efficiency. Cleaner documentation supports more accurate billing, fewer reimbursement delays, and clearer insights into system performance. It also enhances workforce morale by addressing one of the most persistent sources of frustration: documentation fatigue.

This is not about reducing the role of the EMS clinician. It is about amplifying it, ensuring that the decisions made in chaotic, high-pressure environments are captured, defended confidently, and connected seamlessly to the broader healthcare system.

This is the win-win-win: better care, better protection, better outcomes.

The Clinician's Duty Does Not Disappear

AI may be transforming how EMS documentation is created, but it does not change who is ultimately responsible for the record. No matter how sophisticated the tool, the EMS clinician remains the author of record, the witness to events, and the final decision-maker. AI can accelerate the process, reduce clerical burdens, and enhance consistency, but cannot replace clinical judgment or professional accountability.

- **AI can draft, only the clinician can author:** While an AI system may transcribe events with remarkable fluency, it cannot interpret fear in a parent's voice, the chaos of a hoarder's home, or the intuition that something is wrong. It cannot understand why you chose to extricate before establishing an IV, or why you diverted to a stroke center despite being closer to another facility. These moments require context—context that only the clinician possesses.

- **Accountability begins and ends with the clinician:** Whether a report is typed manually, dictated, or generated through AI assistance, the same standards apply: the documentation must be clinically sound, legally defensible, and professionally complete. If a report becomes the subject of a complaint, investigation, or lawsuit, the presence of AI will not shift responsibility. It is your name on the chart, your license at stake, and your decisions under review.
- **AI must be used to elevate, not abdicate:** When applied thoughtfully, AI can help clinicians structure their thoughts, catch errors, and reduce fatigue. It can provide a starting point—but it cannot be the endpoint. If the tool begins to override the clinician's thinking or reduce engagement with the narrative process, it becomes a liability instead of a safeguard.

Disclosure is part of that accountability. Patients who know an AI tool is being used and have a moment to agree, ask questions, or object, are patients who trust the record that follows. A 12-second verbal disclosure at the start of an encounter ("I'm using an AI tool to help me with my documentation. Is that okay with you?") is an important mechanism to gain trust and embrace transparency. It is the same informed-consent ethos that governs every other meaningful intervention in our profession. Also, state law is moving fast on this point. Texas (TRAIGA, effective January 2026) now requires written disclosure of AI use in diagnosis or treatment.[7] California (AB 489, effective January 2026) prohibits AI systems from implying they are licensed providers. Class-action lawsuits testing these principles under older wiretapping and confidentiality statutes are already in court.[8] Whether any lawsuit succeeds or fails, the direction is clear: Agencies that build disclosure into workflow now will carry the lowest risk and the greatest patient trust, regardless of how the case law settles.

AI does not see the scene, feel the urgency, or make judgment calls under pressure. You do. That reality must remain at the center of EMS documentation, no matter how advanced the technology becomes.

The clinician's duty is grounded in presence—what was seen, heard, decided, and done. Yet for too long, EMS has allowed others to shape that record. Without clinician-controlled footage or audio, our actions are judged through the partial lens of a bystander's phone or a police officer's body cam.

It is time to reclaim that lens. Today's body-worn and in-vehicle cameras are no longer just tools for accountability. They are integral to AI-powered documentation, clinician protection, and real-time decision support.

Leveraging the Lens: Body-Worn and In-Vehicle Cameras in Emergency Medical Services

In high-risk, high-accountability environments, visibility is protection. Law enforcement understood this. They embedded cameras into policy, practice, and culture. EMS did not. Clinicians are more exposed and less equipped to defend their actions, and every year, we delay.

As a former state EMS director, I oversaw the adjudication of hundreds of serious complaint cases—use-of-force allegations, refusals gone wrong, critical airway interventions, sexual misconduct reports, pediatric trauma calls. In more than half of those cases, there was video evidence. But not once—not a single time—did that footage come from EMS. The video always came from someone else: a police officer's body cam, a bystander's phone, a traffic camera, a doorbell feed, or a hospital security system.

This imbalance is dangerous. It means your actions are being judged through someone else's lens—often without sound, without context, and without the dynamic decision-making environment you operated in. It means you are visible but voiceless in the documentation of high-risk encounters.

This is not about staging scenes or protecting bad behavior. This is about ensuring that EMS clinicians have access to the same evidentiary tools every other high-risk profession now considers standard. A correctly implemented body-worn or fixed camera system can

- corroborate accurate documentation,
- protect clinicians against false accusations,
- provide objective evidence in use-of-force or patient interaction disputes,
- enhance training and clinical review with real-world footage, and
- reinforce a culture of professionalism and accountability.

However, none of that happens if EMS systems fail to act.

The 2021 *NEMSIS EMS Body-Worn Camera Quickstart Guide: Legal Considerations* outlines the key legal, ethical, and operational issues you must address.[9] Patient privacy, data retention, informed consent, and state wiretapping laws are all manageable with proper policy design and training. Do not use legal complexity as an excuse for technological inaction. Use it as a roadmap to responsible implementation.

Your perspective matters. In every major clinical or operational dispute, the absence of your footage is a disadvantage. It is not about being on guard or performing for the lens. It is about ensuring the truth of your actions is captured in context—your context.

And the opportunity extends far beyond legal defense.

With current technology, body-worn and in-vehicle cameras are no longer passive recording devices—they are intelligent data sources. The U.S. Department of Justice, in its 2025 AI policy update, explicitly identified the use of video analytics, voice transcription, and contextual tagging from camera systems as core enablers of AI-powered reporting and decision support.[10] In EMS, this means

- voice-to-text field documentation that auto-captures the patient's story and timeline,
- real-time triage and decision support using computer vision and audio cues,
- post-incident review and QA enhanced by structured metadata, and
- AI-assisted alerting for escalating behavioral or clinical signs during encounters.

These tools are not futuristic—they are operational today in other sectors. EMS must stop acting like we are five years behind health care and public safety. The technology exists. The standards exist. The leadership decision is what is missing (fig. 7–4).

Start with high-risk scenarios: violent scenes, custody transfers, behavioral emergencies, and refusals of care. Build policy, protect privacy, and train your team, but stop waiting for perfect conditions. You are already being recorded. The only question is: will your point of view, as the EMT or paramedic, be represented?

Reclaiming the Narrative: Professionalism, Protection, and the Power of Presence

EMS documentation does not begin in an office—or end in a courtroom. It starts at the scene. Amid the chaos of a call, the narrative becomes the clinician's enduring voice: a professional account of what happened, why decisions were made, and how care was delivered in complex, uncontrolled environments.

But in too many systems, that voice remains hidden.

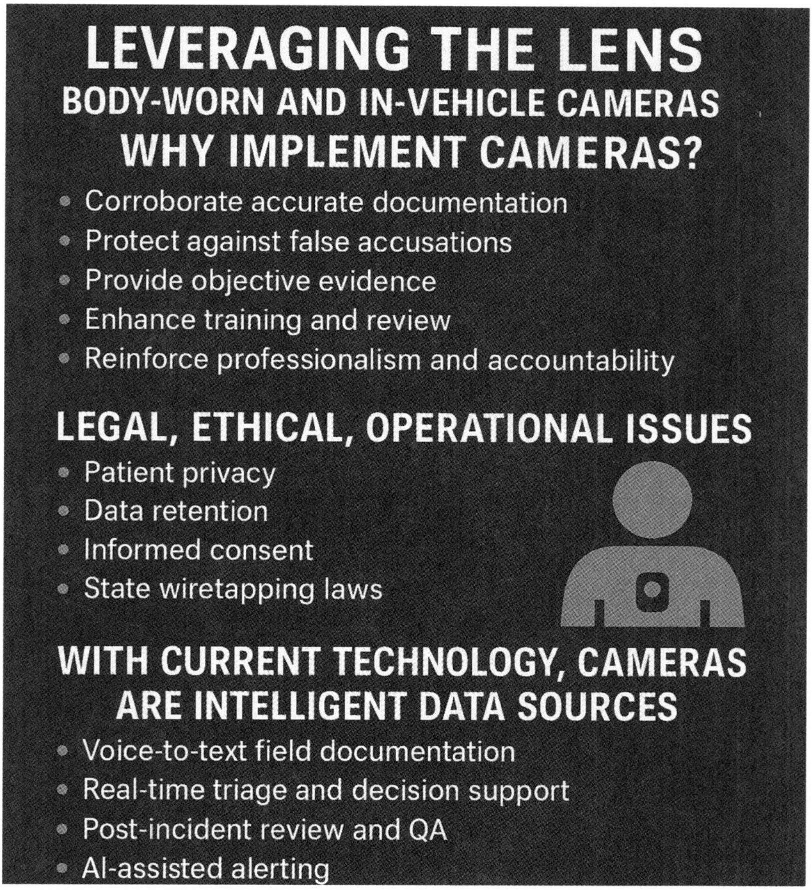

FIGURE 7–4. Cameras are transforming EMS documentation. Beyond accountability and training, modern AI-powered systems can convert audio into structured reports, assist with QA/QI, and strengthen transparency in patient care.

Although EMS generates more than 60 million patient care reports yearly, these narratives are rarely shared with patients, often truncated before reaching hospitals, and frequently siloed from the broader healthcare continuum. Yet these reports contain something no other provider captures: the earliest clinical impressions and the full complexity of the scene. We enter homes no one else sees. We make decisions no algorithm could fully anticipate. And we carry the weight of those choices long after the call ends.

A well-written EMS narrative is more than a timeline. It reflects clinical judgment, emotional clarity, and professional accountability. When charted

Verbalize for the Camera—Document Like You're on Record

With cameras in the ambulance and on your person, your actions are no longer just seen—they're heard. In high-risk calls, this is your shield. Use it wisely. EMTs and paramedics were taught in the now-retired psychomotor exams: *"If you didn't say it, you didn't do it."* While that never matched real-world field practice, the return of verbalized action has new value, not for testing, but for protection.

Clinician Tips for Leveraging Video Devices in Real Time

- **Narrate your general impression.** As you visually assess, say what you see: "Patient appears pale and diaphoretic. Abdomen firm, with tenderness in the lower quadrants." You're not performing for the camera—you're capturing clinical reality in your voice and actions.
- **Use verbal summaries during transitions.** Say it out loud before you move or transfer: "We're moving the patient with suspected pelvic trauma using a scoop stretcher and a pelvic binder. The airway is patent, and the patient is alert."
- **Announce key findings and rationale.** Verbalize vital signs, concerns about changes, or your reason for choosing one hospital over another. "We're diverting to the stroke center due to unilateral facial droop and slurred speech."
- **Treat video as part of the record, not a risk.** You're not creating evidence for others. You're building your narrative in context, for protection, clarity, and clinical documentation.

The Health Insurance Portability and Accountability Act and Video Storage: Don't Let It Be an Excuse

EMS leaders sometimes cite HIPAA as a barrier to adopting video systems; however, it is not. HIPAA allows for the collection, storage, and use of audio, video, and written documentation, as long as systems meet privacy, access control, and retention requirements:

- Video files can be encrypted and stored just like ePCRs.
- Audio and visual data can be managed under the same HIPAA rules that apply to radio transmissions, telemetry, and mobile apps.
- Cloud-based storage with access logging and role-based permissions is compliant and widely used across health care.

HIPAA is not a barrier—it is a standard. One EMS must meet, not avoid.

Narrate when it counts.

Speak like your voice matters—because it does. When the footage is reviewed by a medical director, a supervisor, or a courtroom, your voice should tell the story, not someone else's hidden camera or cell phone footage from across the street.

Note also that HIPAA is the federal floor, not the only law in play. In two-party consent states, including California, Florida, Illinois, Pennsylvania, Washington, and a dozen others, audio recording of a clinical encounter may require all parties to be informed and to consent, regardless of HIPAA.[11] Recent class-action litigation against health systems deploying ambient AI scribes has been filed under these older wiretapping statutes, not HIPAA. Agency counsel should review state law before deployment.

with care and completeness, it protects the clinician, supports hospital care, and advances patient understanding. When vague or incomplete, it invites confusion, erodes trust, and increases the risk of legal or disciplinary action. As a former state EMS director, I have seen both outcomes—strong narratives that close investigations quickly, and weak ones that trigger unnecessary harm.

Fair or not, the written word often shapes the outcome. Your electronic signature is not just a formality. It is a legal declaration. It says: I was there. I understood. I acted with purpose. AI cannot replace that presence, but it can support it.

Used wisely, AI can prompt for missing details, reduce cognitive burden, and help clinicians produce clear, timely, and context-rich documentation. It can transform charting from a disconnected compliance task into an integrated communication tool—accessible, shareable, and clinically relevant.

This is not just about better documentation. It is about professional visibility. It is about redefining EMS not as "just the transport," but as the first clinical contact—the point of entry where stories begin and critical decisions are made.

If we do not document it, that part of the story disappears. And if we do not tell it, no one else will.

AI gives us the means, professionalism provides us with the mandate, and leadership gives us the path.

Now is the time to reclaim the EMS narrative—and ensure it is written, shared, and heard.

Smart Narratives: Where Artificial Intelligence Meets Clinical Judgment

AI can generate fluent sentences, but it cannot walk through the scene. It does not smell the smoke, feel the tension, or hear the urgency in a parent's voice. EMS clinicians do. That is why AI cannot replace the patient care narrative. But when appropriately designed, it can support it.

The next generation of documentation tools must go beyond transcription. They must become narrative aware, capable of recording speech, and understand EMS care's cadence, logic, and structure. These tools must recognize clinical reasoning, highlight inconsistencies, and reinforce best practices—not just in what happened but also in how and why decisions were made.

These are smart narratives—reports coproduced by human clinicians and intelligent systems. In a smart narrative, AI becomes more than a scribe. It becomes an editorial partner that helps structure thought, reinforce clarity, and anticipate gaps (fig. 7–5).

For example, a smart documentation system can do the following:

- **Prompt** for missing elements when symptoms are mentioned but not explored.
- **Assist** with structuring the sequence of events so the report flows logically and chronologically.
- **Highlight** contradictory or vague phrases that could lead to regulatory scrutiny.
- **Flag** high-risk refusal or delayed intervention scenarios for immediate QA review.

As mentioned in the chapter introduction, this approach does not diminish the clinician's role—it elevates it. The clinician remains the final author. AI can suggest but not certify. It can compose but not witness. The narrative—both as legal evidence and professional expression—must reflect the person's judgment there.

The promise of smart narratives is not that AI will tell our stories better than we can. The promise is that AI will help EMS clinicians tell their stories more clearly, completely, and confidently than ever before. It will also reduce documentation times, improve completion rates, and minimize one of the most persistent stressors for EMTs and paramedics. Let us be honest: No one enters EMS for the love of charting.

Smart narratives can ensure that documentation is consistently legible, professionally structured, and clinically accurate. They also allow clinicians to capture better the details that often get lost—what the scene looked like, what it smelled like, the obstacles in the hallway, the shouting, the confusion, the clutter. These elements define EMS care and distinguish it from every other healthcare environment. AI should not erase that reality. It should help preserve it.

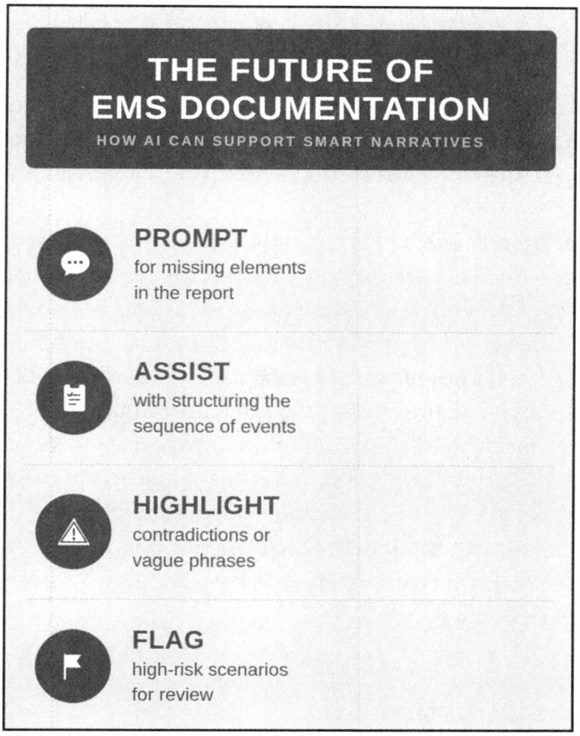

FIGURE 7–5. AI can strengthen documentation quality by prompting for missing data, organizing events chronologically, and flagging inconsistencies. Smart narrative systems will enable clinicians to tell the full clinical story accurately and efficiently.

Key Terms and Definitions

- **Patient Care Narrative:** A free-text, clinician-written description of the EMS encounter, documenting context, scene dynamics, decision-making, and patient interactions. It is contemporaneous legal evidence and a critical component of EMS documentation.
- **Structured Data:** Discrete data elements (e.g., time stamps, vitals, interventions) entered into standardized fields in an ePCR, enabling interoperability, surveillance, and large-scale analytics.
- **Large Language Model (LLM):** An advanced AI system trained on vast datasets to understand, generate, and summarize human language. In EMS, LLMs assist in composing or structuring patient care narratives.
- **Voice-to-Narrative Technology:** A tool that allows EMS clinicians to speak naturally while AI captures and converts the speech into structured, readable, and editable narrative documentation.
- **NEMSIS (National EMS Information System):** The national EMS data repository managed by the National Highway Traffic Safety Administration that captures over 60 million EMS activations annually, enabling public health surveillance, policy development, and operational analysis.
- **Documentation Burden:** The cognitive, time, and emotional workload associated with completing patient care documentation, particularly narratives and structured ePCR fields, is often cited as a source of burnout.
- **Technical Debt:** The legacy software limitations that hinder the adoption of modern innovations, such as AI integration, due to outdated code architecture or proprietary design constraints.
- **Narrative Integrity:** The preservation of human-centered, situational context in EMS documentation, ensuring the narrative reflects real-world complexity and clinician decision-making.
- **Contemporaneous Evidence:** Documentation written at or near the time of an event, carrying substantial legal weight in administrative, legal, or regulatory reviews.

Reflection Questions

1. When was your organization last reviewed on the legal importance of EMS documentation with your clinicians? How is "contemporaneous evidence" reinforced in training and QA?

2. How would AI-powered documentation improve—or challenge—your agency's current documentation workflows? Where do you see resistance, and how can you address it?
3. What safeguards or review processes should be in place to ensure AI-generated narratives maintain clinical and legal accuracy? Who is ultimately accountable for what is submitted?
4. How can your agency use documentation innovation to better integrate with hospital systems, engage patients, and highlight EMS's value in the healthcare chain? What needs to change in how we share and access EMS reports?
5. If you were designing an EMS narrative from scratch using AI tools, what elements must be captured to protect your clinicians and reflect the reality of the scene? Are those elements currently being documented?

Study Questions

1. Which of the following is not a reason EMS documentation has historically been underutilized?

    a. Minimal training in narrative writing
    b. Time pressure during long transports
    c. High investment in AI-powered systems
    d. Lack of consistent emphasis on documentation quality

2. True / False: Structured data fields alone are sufficient to capture the clinical decision-making of an EMS call.
3. AI can draft. Only the _________ can author.
4. Describe why real-time verbalization of findings is important in a video-recorded EMS environment.
5. According to the 2021 NEMSIS Guide, which of the following is not a legal consideration when implementing camera systems in EMS?

    a. Patient privacy
    b. Informed consent
    c. Crew scheduling
    d. State wiretapping laws

6. True / False: HIPAA prohibits the use of audio and video recordings in EMS documentation.
7. Body-worn and in-vehicle cameras are no longer passive recording devices—they are _________ data sources.

CHAPTER SUMMARY

Artificial Intelligence–Powered Documentation—Reclaiming Time and Protecting Integrity

For over 50 years, EMS has led the nation in real-time healthcare data collection. From Dr. Peter Safar's 1969 call for standardized reports to today's national NEMSIS system, which captures 58 million activations a year, EMS pioneered structured data systems long before the rest of health care caught on.

However, structure alone is not enough.

The patient care narrative remains one of the most powerful tools we have. It defends clinicians, documents context, and captures the complexity of care in environments no other medical professional encounters—chaotic homes, icy ditches, loud highways, and dangerous scenes. That narrative is more than a paragraph. It is contemporaneous evidence, written at the moment, long before any complaint or litigation.

In 2023, the EMS profession entered the next chapter: AI-powered documentation. While initially dismissed, the concept seeded a revolution. LLMs now assist in capturing EMS narratives with unprecedented accuracy, speed, and structure, giving time back to clinicians while elevating the quality of the report. And EMS is not just catching up—we led the way.

By 2025, major ePCR vendors had begun integrating AI tools: voice-to-narrative engines, smart QA prompts, and real-time clinical reminders. The promise is clear, but the responsibility remains. AI can draft, but only EMS clinicians can decide what stays in the chart.

This chapter reminds us that leveraging AI in EMS documentation is not just about efficiency. It is about professional integrity, legal protection, and reclaiming ownership of our voice in health care.

The future of AI-powered EMS documentation is not just about lightening the load or saving time (though that is a bonus!); it is about innovative solutions that enhance patient outcomes, bolster national security, and showcase the essential role of EMS in health care—and we have the power to write our narrative!

8

Real-Time Decision Support

LEARNING OBJECTIVES

After reading this chapter, the reader will be able to

- Describe the operational challenges EMS clinicians face in making high-stakes decisions under real-world, low-resource conditions.
- Differentiate between traditional EMS decision aids (e.g., protocol books, pocket cards) and AI-driven real-time decision support tools.
- Explain the clinical potential of AI systems, such as sepsis alerts, modified shock index tracking, and augmented reality overlays, to enhance field diagnostics, improve response time, and reduce error.
- Identify existing real-time decision support models in health care, such as Epic's Sepsis Prediction Model and Ambient's Sepsis Detection and Response Tool™, and evaluate how similar systems can be translated to EMS settings.
- Analyze how AI-powered systems can reduce ambulance patient offload times, support operational decision-making, and foster EMS-hospital coordination using shared data.
- Evaluate the ethical, legal, and regulatory implications of integrating AI into field decision-making, especially regarding FDA oversight and clinical accountability.
- Articulate the potential for real-time AI to improve care equity by scaling expert clinical support to under-resourced, rural, or volunteer EMS systems.
- Reflect on EMS leadership's role in driving the responsible adoption of AI tools, ensuring clinician training, policy alignment, and equitable implementation.

Prologue: Decisions at Sixty Miles Per Hour

You are nearing the end of an exhausting shift, transporting what you hope will be your last patient. He is a 65-year-old male with chest pain following a low-speed car crash. He is pale, diaphoretic, slightly confused, and clutching his chest. His electrocardiogram (ECG) is borderline. His blood sugar reads 112. His med list includes Jardiance, Ozempic, a beta-blocker, and one you have never seen before.

You've seen patients like this before. But this time, your intuition is screaming that something is different.

As you interview the patient, an embedded microphone records the conversation, like in your primary care physician's office. The AI engine transcribes and analyzes the exchange in real time, flagging inconsistencies between what the patient says and what the data shows. Your monitor streams vitals directly into your onboard system—heart rate, blood pressure, oxygen saturation, 12-lead—all feeding a live decision support algorithm.

Your AI display flags the unfamiliar medication. It's a newly approved anticoagulant—and that detail now matters.

Then comes the alert.

Before your eyes even register the change, your heads-up display flashes a warning:

"Modified shock index rising: 1.0 → 1.5. The pattern suggests an early hemorrhagic compromise."

The patient is decompensating, and the system saw it first.

Your smart glasses now project a soft overlay across the patient's chest—visible only to you—highlighting subtle asymmetries in thoracic movement. Facial video sensors, backed by LiDAR (light detection and ranging) and biometric AI, detect skin tone changes and respiratory shifts you might have missed. This is real-time clinical augmentation, and all the technology and components are already being tested; it is not theoretical.

This is not science fiction. It is the future of EMS. And it's already in service.

The technology is silent, seamless, and clinician-first. You remain in charge. The system does not override your judgment—it augments it. However, your decisions are now faster, sharper, and backed by data you cannot access alone.

You do not miss the modified shock index (MSI) shift. You do not delay the transport decision. You are not left guessing or catching up. You are ahead.

And that is precisely where EMS belongs.

EMS clinicians face a profound disadvantage compared to our hospital-based peers. We work alone or in pairs, in uncontrolled environments, performing every role—interviewer, diagnostician, proceduralist, medical calculator, documentarian, and educator. We often lack access to prior medical records, are forced to decipher handwritten med lists while bouncing down highways, and are expected to make split-second decisions with incomplete information.

Meanwhile, in Sri Lanka, EMS clinicians have already deployed a connected ambulance model powered by real-time AI overlays, telemedicine access, and augmented reality interfaces. Since 2023, this system has reduced cognitive load, expanded care teams, and provided virtual backup on demand. With Microsoft HoloLens-enabled headsets, clinicians receive live support from remote physicians, with real-time vitals, ECGs, and interactive overlays to guide care, all on low-cost platforms built for scalability.

It is already happening.

Hospitals now use predictive models, and primary care physicians review AI-generated summaries before patients leave. However, despite being the literal front door of health care, EMS is still taping paper med lists to stretchers and documenting long after the call ends.

That is no longer acceptable either.

This chapter is about what is possible, what is already real, and how EMS can lead this innovation in medicine.

When responsibly integrated, AI will not replace the EMS clinician but amplify them. It will restore bandwidth, reduce errors, enhance safety, and make every response smarter than the one before.

While decision support tools at the point of care receive much attention, real-time augmentation must begin before a clinician arrives on the scene. Real-time intelligence can have the most immediate and lifesaving impact at the first link in the EMS chain, dispatch.

The Evolution of Decision Support in Emergency Medical Services

As the EMS profession matured, patient acuity increased, medication lists grew longer, and care shifted toward older, more complex populations, the cognitive demands placed on EMS clinicians expanded beyond what static guidance could reasonably support. What once fit within a simple, pocket-sized protocol reference now requires integrating physiology, pharmacology, context, and patient trajectory.

From Protocol Books and Pocket Cards to Predictive Intelligence

Field medicine is one of the most cognitively demanding arenas in health care. EMS clinicians are expected to assess, treat, calculate, document, and transport—often simultaneously—while navigating high-pressure environments, limited information, and unpredictable conditions. There is no pause button, clean room, and often no second opinion.

EMS professionals have relied on laminated cards, printed protocols, and personal memory to guide their decisions for decades. These tools served a critical role. They provided structure and consistency, but they offered little adaptability or foresight. They supported action but not anticipation.

That is no longer enough.

A New Era of Support

The evolution of decision support in EMS mirrors the transformation across health care: from static tools to dynamic, real-time guidance.

Those large protocol books—and I still have copies of mine from 20 years ago—have been digitized into protocol apps or integrated into electronic patient care reporting (ePCR) systems. While this integrated approach has added value—the drug calculators are far better than my manual calculations at 2 a.m.—they still operate within the narrow logic of "if this, then that." The current systems do not adapt to the nuances of each patient, let alone anticipate what might happen next.

Today, we have the opportunity to go further.

Artificial Intelligence–Powered Decision Support

Flowcharts no longer bind the next generation of clinical decision support tools. They can ingest live data—vital signs, medications, mechanism of injury, scene inputs—and provide clinicians with timely, context-specific insights. These tools do not replace your clinical judgment—they sharpen it (fig. 8–1). For example,

- An AI engine flags a rising MSI before you detect clinical deterioration
- A subtle change in skin tone, captured on camera, prompts early recognition of shock or cyanosis
- A medication profile cross-referenced with vital signs alerts you to an underappreciated drug interaction

The technology now exists and will soon intersect with EMS. It is not about making us dependent—it is about *making us stronger*. But perhaps the most profound evolution will be in how AI enables truly personalized clinical intelligence for individual patients.

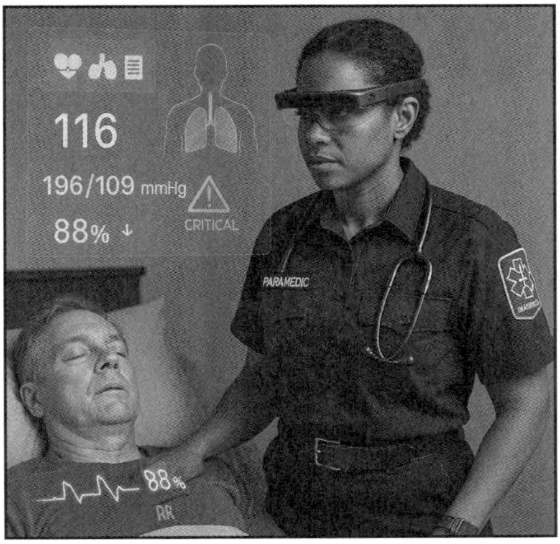

FIGURE 8–1. AI-driven augmented reality provides paramedics with continuous assessment cues, integrating vitals, alerts, and predictive analytics directly into their line of sight. This fusion of data and visualization represents the next stage of evolution in prehospital care. (Photo is AI-generated image.)

The next generation of decision support will integrate patient-specific AI agents with health information exchanges, allowing EMTs and paramedics to access synthesized clinical intelligence in *real time*. Consider responding to a syncopal episode in an 82-year-old patient with multiple comorbidities. Rather than relying on the patient (or family) for medical history, scrolling through fragmented records across multiple health systems—or having no access at all—the EMS clinician will be able to voice query a HIPAA-compliant AI system that has full access to the patient's records. While treating the patient, the clinician can ask the AI model: 'What medications changed in the last 30 days?' or 'Has this patient had similar episodes?' The system would analyze the patient's complete medical history and provide relevant clinical insights, highlight recent medication or diagnostic changes, flag abnormal lab results, and surface patterns pertinent to the current presentation. Early versions of these systems are already being tested in pilot programs.

This capability represents a fundamental shift in how field medicine operates. EMS clinicians will leverage this technology to deliver better, more relevant care customized to each patient while saving tremendous time currently lost to information archaeology. The technical infrastructure already exists, including health information exchanges, HIPAA-compliant AI frameworks, and consent management systems that are operational. What's required is intentional integration. These new software solutions will not replace clinical judgment, but will collapse the gap between question and insight, allowing paramedics to focus cognitive energy where it belongs: on the patient in front of them.

EMTs and paramedics make decisions under pressure, and the more credible information you have, the faster and safer you move. Real-time support helps reduce diagnostic error, standardize care across systems, and enhance patient safety—even when flying solo in the back of a moving rig.

The next generation of AI will also open new doors: more intelligent quality assurance, stronger documentation, faster turnarounds, and fewer missed red flags.

Artificial Intelligence in Emergency Medical Services Dispatch: From Call-Taking to Real-Time Intelligence

Real-time decision support in EMS does not begin at the patient's side; it starts when the call is placed. Across the United States and around the world, 9-1-1 centers are under increasing strain. Call volumes continue to rise, staffing levels have fallen, and emergencies have grown more complex. In this high-pressure environment, AI is no longer theoretical. It is reshaping how EMS systems assess, triage, and respond—beginning with dispatch.

The modern 9-1-1 system is stretched thin. National data show vacancy rates nearing 25% in communications centers across all 50 states. Accidental calls—many triggered by smartphones and wearable devices—now make up a growing share of volume, consuming time and resources with little clinical value. At the same time, real emergencies demand faster triage, better information, and clearer coordination.

Telecommunicators remain the first point of contact in most EMS activations. They manage simultaneous demands: entering data, coordinating multiple units, listening for critical cues, and calming distressed callers—often while navigating up to four software systems simultaneously. This environment involves rapid decisions and high cognitive load. Even small inefficiencies can have life-or-death consequences.

AI tools are not replacing this essential role. They augment it, and the results change what dispatch means for clinicians, patients, and systems under pressure.

A System Under Pressure

America's 9-1-1 infrastructure is facing its own emergency. In 2023, a joint national survey[1] conducted by the International Academies of Emergency

Dispatch and the National Association of State 911 Administrators revealed that more than one-third of dispatch centers were operating with vacancy rates above 30%. Some reported staffing gaps exceeding 80%. Nearly 4,000 trained professionals exited the profession in a year, citing pay, stress, shift demands, and emotional burnout as driving factors.

This workforce shortage comes amid rising call volume. Smartphones, smartwatches, and voice-activated assistants now trigger thousands of accidental calls, further overwhelming already short-staffed centers. Each unintended ring diverts attention from actual emergencies and erodes the system's ability to respond quickly.

Behind every headset is a telecommunicator carrying the extraordinary cognitive load. Dispatchers juggle real-time call intake, multiscreen data entry, incident updates, responder coordination, and emotionally distressed callers—often simultaneously. These professionals are the first link in the EMS chain, and their performance directly influences scene coordination, response times, and ultimately, patient outcomes.

In this environment, even 10 seconds matter. A missed detail, misunderstood address, or delayed tone-out can mean the difference between life and loss. The current system is not just strained—it is unsustainable. Without urgent innovation, the gap between call volume and call handling capacity will widen.

AI offers a path forward—not to replace dispatchers but to support them, enhancing accuracy, improving speed, and preserving capacity, one of the most critical functions in emergency response.

What Artificial Intelligence Dispatch Is

AI in EMS dispatch is not a replacement for human telecommunicators—it is a force multiplier. These systems use a combination of machine learning, natural language processing, real-time analytics, and predictive modeling to enhance performance in high-stress environments. AI is not taking over the console—it is listening, analyzing, assisting, and prompting in ways that help dispatchers stay ahead of demand and reduce risk under pressure.

Key capabilities of AI-enabled dispatch systems include the following:

- **Real-time transcription and insight extraction** from caller conversations
- **Live translation across dozens of languages,** supporting linguistically diverse communities
- **Voice stress and urgency detection** using tone analysis and natural language processing to surface hidden distress
- **Duplicate call clustering** to identify single incidents from multiple 9-1-1 calls during mass events

- **Predictive modeling of call volume and unit demand,** guiding dynamic resource positioning
- **Smart triage prompts and deployment recommendations** based on historical and real-time data patterns

These tools can detect a whispered threat buried in background noise, recognize that five separate calls describe the same crash, and recommend the optimal hospital destination based not just on geography but also on real-time data about emergency department crowding and specialty availability.

More importantly, AI enhances the accuracy and clarity of information relayed to field crews. Cleaner dispatch data supports faster scene understanding, better clinical preparedness, and safer patient care. In the high-stakes seconds that define EMS response, that kind of intelligence is not optional—it is essential.

Case Studies in Dispatch Innovation

Across the globe, EMS systems are already demonstrating how AI can strengthen dispatch operations and improve outcomes. In Grand Rapids, Michigan, Life EMS Ambulance implemented an AI-enabled system in 2024 known as Predictive Dynamic Deployment. By analyzing historical call data, weather patterns, traffic congestion, and geographic risk zones, the platform forecasts where incidents are most likely to occur. Ambulances are then repositioned in real time, allowing units to be closer to likely events before they happen—the result: improved response times and more efficient system utilization across the region.

In South Korea,[2] the Intelligent Emergency Medical Service system—developed through a national 5G-based AI initiative—demonstrates what fully integrated emergency response can achieve. Spearheaded by Severance Hospital and supported by the Ministry of Science and the Ministry of Health and Welfare, the system represents a nationwide effort to optimize every stage of emergency care. It leverages 5G connectivity, cloud infrastructure, and real-time AI analysis to classify patient severity, recommend standardized first aid, and guide ambulances to the most appropriate hospital based on live capacity and specialty data. This system actively supports first responders by collecting and analyzing 360-degree video, bio-signals, and audio data to generate emergency records and coordinate hospital handoff—all while in transit. Targeting Korea's four most critical emergency conditions—cardiac arrest, cardiovascular and cerebrovascular events, and severe trauma—the system has already demonstrated improvements in timeliness and care coordination, securing the golden hour and aiming to significantly increase survival rates. It has been deployed, publicly demonstrated, and funded as a national priority.

Several platforms are already used across U.S. dispatch centers, including Prepared Assist, Corti Engage, Carbyne, and HighWind AI. These tools bring automation to previously manual tasks. They transcribe audio and text in real time, translate across dozens of languages, highlight caller location and contextual keywords, and consolidate workflows onto a unified interface. This capacity is more than convenience in high-risk situations—it is lifesaving.

In Delaware County, Pennsylvania, a whispered phrase—"he pointed a gun at me"—was nearly lost in the chaos of a 9-1-1 call.[3] But the system's transcription engine detected the statement, flagged it for review, and alerted the dispatcher. The call was escalated immediately, and the response likely prevented a tragedy.

These are not abstract promises. They are examples of what happens when AI is applied purposefully, designed to enhance—not replace—the humans behind the screen.

Potential to Reshape Emergency Medical Services Systems

AI-enabled dispatch can fundamentally improve EMS operations when implemented with care and oversight. These tools do not replace telecommunicators' human judgment—they *augment* it. By automating repetitive tasks, enhancing situational awareness, and surfacing insights in real time, AI can help dispatchers and clinicians focus on what matters most: effective, timely patient care.

Strategically deployed, AI dispatch systems can do the following:

- **Improve response times** through predictive deployment and dynamic routing based on real-time conditions
- **Reduce dispatcher workload** by automating manual inputs and assisting with transcription, translation, and call clustering
- **Enhance situational awareness** for EMS crews through more complete, accurate, and context-rich information before arrival
- **Support alternative responses** for behavioral health, substance use, and social crisis calls, enabling triage to more appropriate resources
- **Reduce unnecessary responses** and better match clinical units to patient needs by leveraging data-driven acuity assessments
- **Advance equity in emergency care** by minimizing bias in call triage and interpretation, especially for non-English speakers and marginalized populations

These capabilities also unlock new potential for system-wide analytics. Trends in call types, service demand, and regional public health can be monitored in near real time, allowing EMS leaders to shift from reactive deployment to proactive forecasting. With proper governance, AI dispatch is not just a technical upgrade but a strategic tool for building smarter, safer, and more responsive EMS systems.

The clinical potential of AI-assisted dispatch is already being demonstrated. A 2021 study published in *Resuscitation*[4] evaluated how a machine learning model could assist human dispatchers in recognizing out-of-hospital cardiac arrest. The results were striking: The machine learning tool identified out-of-hospital cardiac arrest within the first minute in 36% of calls, compared to 25% for dispatchers. Across the call duration, machine learning achieved a median recognition time of 72 seconds versus 94 seconds for human telecommunicators—a 22-second improvement in a scenario where every second matters. These findings suggest that AI-supported dispatch systems can significantly improve time-critical care identification, especially in high-stakes, high-volume environments.

Ethical, Clinical, and Operational Considerations

AI-enabled dispatch systems offer enormous potential, but they also introduce new risks. Like any tool used in clinical or operational decision-making, clear guardrails must govern these systems. Without strong oversight, AI can inadvertently reinforce inequities, misclassify calls, or compromise patient trust.

Common concerns include the following:

- **Bias in triage prioritization,** particularly for historically marginalized or underrepresented populations
- **Insufficient oversight** of training data, thresholds, or alert algorithms that influence triage accuracy
- **Overreliance on automation,** where human judgment is sidelined or dulled over time
- **Data privacy vulnerabilities exist,** especially in cloud-based systems that transcribe or analyze real-time audio streams

These concerns are not hypothetical. They are already showing up in broader healthcare AI use, and EMS cannot afford to ignore them.

Accountability remains with the telecommunicator. Medical oversight must govern how triage decisions are made, what data is prioritized, and how alerts are escalated. EMS agencies must ensure that AI tools are transparent in design, explainable in action, and supervised by qualified professionals. Patient

CASE STUDY

Automation in Emergency Medical Services Improves Outcomes

While many real-time decision tools in EMS are still emerging, evidence from international systems already shows what is possible when automation is thoughtfully implemented. A 2021 study[10] published in *BMC Medical Informatics and Decision Making* evaluated a comprehensive EMS automation system developed at Kerman University of Medical Sciences in Iran. The goal was to improve prehospital care timeliness, coordination, and quality through automated call handling, patient tracking, and clinical data sharing with hospitals.

The results were striking:

- Average hospital stays dropped by 3 hours and 45 minutes.
- Emergency department discharges increased by 2.2%.
- Patient mortality was reduced by 1.3%.

The automation system also improved workflow for EMS personnel and hospital staff alike. More than 75% of dispatchers, EMTs, nurses, and physicians reported increased speed, better coordination, and improved decision-making. The study concluded that EMS automation systems could significantly enhance patient outcomes and workforce efficiency, especially when paired with real-time telemetry, hospital integration, and decision support tools.

While infrastructure gaps and regulatory limitations posed challenges, this research demonstrates that AI-enabled systems are beneficial and transformative. For EMS leaders, the takeaway is clear: real-time automation is no longer experimental. It is a proven driver of better outcomes, reduced delays, and more empowered clinicians.

Some readers may be unfamiliar with EMS systems outside the United States or hesitant to draw lessons from international case studies. That hesitation is understandable, but it should not preclude learning. While conducted in Iran, the *Afzali et al.* study was published in a well-respected, peer-reviewed journal (*BMC Medical Informatics and Decision Making*) and follows a structured, quantitative methodology that aligns with global research standards. More importantly, the system challenges this study addresses—workflow

(continues)

> inefficiency, fragmented communication, and hospital offload delays—are like the challenges encountered in U.S.-based EMS. While the specific infrastructure or regulatory context may differ, the core principles of automation, integration, and real-time visibility apply across borders. The study should not be viewed as a prescriptive model, but as evidence that even basic automation can yield measurable clinical and operational gains when designed around frontline needs.

data—spoken, typed, or inferred—must be protected with the same rigor as any electronic health record.

Transparency, explainability, and human-in-the-loop design must be foundational. Dispatchers must remain accountable. Medical oversight must govern triage logic. Patient data must be protected with the same rigor as any clinical system.

However, these systems are about managing risk and unlocking potential. AI is emerging as one of the few scalable solutions to the EMS workforce crisis. It is not a substitute for human judgment, but a strategic tool to reduce overload, extend capability, and support a profession under pressure. As call volumes rise, response models evolve, and staffing challenges deepen, AI-enabled dispatch will be a critical part of how EMS adapts—not by replacing people, but by empowering them. This begins a broader transformation that reaches beyond the communication and dispatch center.

Artificial Intelligence–Enabled Translation and Language Detection

Language has long been a barrier to accurate assessment and timely care in EMS. In the 1990s and early 2000s, EMS agencies often relied on LanguageLine interpretation services—telephone-based translators that could assist over the radio or phone. These tools served a critical purpose, but they came with significant limitations. Clinicians had to guess the language being spoken to initiate translation, and the service offered no help in understanding written documents or field-specific dialects. In time-sensitive calls, this added friction and delay.

Today, that limitation is gone.

Modern AI tools can detect spoken language in real time and offer instant speech and text translation. Large language models, paired with voice recognition and neural machine translation, can identify over 100 languages, detect code-switching in mixed language households, and provide context-aware translation tailored to clinical settings. EMS clinicians are no longer left guessing. And they are no longer alone.

Imagine arriving on the scene to find a distressed patient speaking Pashto or Tagalog, holding a medical history document in their native script. In 2025, an AI-enabled translation app can scan the document, recognize the language, extract clinical terms, and provide a translated summary—all within seconds. At the same time, real-time speech translation enables dialogue, confirms medication use, and ensures the patient understands the care provided.

This is not just a technical achievement. It is a force multiplier for safety, equity, and professionalism. It reduces risk, enhances communication, and preserves patient dignity, especially for vulnerable populations who may not feel safe speaking in a second language during an emergency.

All EMS systems should adopt this resource, and these AI translation tools should be integrated into ePCR platforms, command center communications, and mobile devices. Leveraging and integrating this technology is another example of how technology will assist with providing clinically sound, culturally competent care—every time, in every language.

What Real-Time Artificial Intelligence Looks Like

To many, *real-time decision support* still sounds futuristic—something reserved for academic centers or research hospitals. However, that future is already here. Epic Sepsis Model Inpatient Predictive Analytic Tool is a case in point.[5]

This embedded system within Epic's electronic health record platform continuously monitors patient vitals, lab values, and medical history. Its predictive algorithms detect early signs of sepsis—patterns that might otherwise go unnoticed—and push real-time alerts directly into the clinician's workflow. The results are profound. A 2023 implementation at a large academic trauma center showed a 44% reduction in the odds of sepsis-related mortality, faster antibiotic administration, increased intensive care unit admissions for appropriate patients, and greater adherence to sepsis protocols.[6]

Epic is not alone. Ambient Clinical Analytics' Sepsis Detection and Response Tool™ offers similar real-time decision support. It provides configurable sepsis alerts, treatment milestone tracking, and centralized data correlation around "time zero" when sepsis is first suspected. These alerts are also pushed to the care team, ensuring coordinated, error-reducing interventions.

These are not pilot projects. They are Food and Drug Administration (FDA)-approved clinical decision support systems that are operational and saving lives.

Emergency Medical Services Implications

Now, imagine this capability, optimized for the field.

You are dispatched to a home for a 72-year-old male with flu-like symptoms: mild fever, borderline tachycardia, and BP in the low 90s. The patient appears fatigued but alert. In a traditional workflow, you might treat this as low acuity. But your real-time AI platform is already evaluating the trajectory in a future-ready EMS system.

Behind the scenes, it's ingesting vitals, calculating an MSI, comparing heart rate variability against decompensation curves, tracking temp drifts, and factoring in the patient's medication profile, which includes Jardiance, a potential flag for euglycemic diabetic ketoacidosis.

Your tablet or augmented reality interface pings discreetly:

> **SEPSIS RISK IDENTIFIED**
>
> Suggest: Early IV fluids. Notify receiving facility: *SEPSIS ALERT.*
>
> MSI rising: 1.0 → 1.5. $EtCO_2$ trending <26 mmHg.

You have not completed your primary assessment, and the AI has spotted the downward turn.

But this system does not replace your decision—it *refines* it. It provides clarity when time is tight and bandwidth is low.

The Clinical Proof: Emergency Medical Services Makes a Difference

Research proves that EMS action—or inaction—has a measurable impact on sepsis outcomes:

- Over 40% of patients hospitalized with severe sepsis arrive via EMS.[7]
- EMS-identified sepsis cases receive antibiotics 52 minutes faster and reach goal-directed therapy nearly an hour earlier.[8]
- Even simple interventions—IV placement or fluid bolus—are linked to 30–50% reductions in hospital mortality.[9]

So why are we not doing this everywhere?

Hospital systems are proving daily that machine-speed analytics, embedded directly into clinical workflows, drive better care and save lives. I would assert that EMS has greater urgency and more significant stakes.

Field clinicians frequently work alone or in pairs, lacking access to labs, prior records, or specialists. The first 10 minutes establish the trajectory, yet the data is limited, and the margin for error is extremely narrow.

That is where real-time AI fits. This is another example of how AI does not replace a clinician's experience or judgment—it's about giving that judgment more power, vision, and foresight.

When seconds matter, decisions need to be better, not just faster.

CASE STUDY

The Sri Lanka Connected Ambulance

Sri Lanka did not wait for someone else to show the way. They built it.

In 2023, Sri Lanka's national EMS system, 1990 Suwa Seriya, launched one of the world's most technologically advanced ambulance programs. This was not born in Silicon Valley or inside a U.S. think tank. It came from a small island nation, using innovative partnerships, political commitment, and a clinician-first mindset to leapfrog the technological curve.

Partnering with Wavenet Global, a Sri Lankan-owned, Singapore-based health technology company, Suwa Seriya integrated AI-driven dispatch, real-time telemetry, and augmented reality overlays via Microsoft HoloLens. The result: a fully digitized, AI-enabled connected ambulance system that closes the gap between prehospital assessment and definitive care (fig. 8-2).

What that means operationally is that EMTs are not working alone anymore. As they manage the patient, the Microsoft HoloLens headset streams vitals, scene visuals, and treatment data to a national command center. Physicians can see what the crew sees, offer

(*continues*)

clinical guidance in real time, and project overlays onto the headset display—turning every response into a coordinated, multiclinician effort before the ambulance arrives.

This is not just about gadgets. It is advancing clinical scope and expectations. EMTs in Sri Lanka are now performing advanced-level interventions under live guidance, without adding a single new full-time specialist to the response model. That is the power of scalable, brilliant innovation.

AI and augmented reality are not luxuries. They are lifelines—especially in under-resourced systems. If Sri Lanka can deploy this technology across a national EMS system amid infrastructure and budget limitations, what's stopping wealthier systems from doing the same?

The answer is not funding. It is vision, motivation, and determination.

If a nation still emerging from decades of conflict can reinvent emergency care with bold leadership and forward-thinking design, then any EMS system can. And if we choose not to? Then we have chosen to let someone else define the future of our profession.

Moreover, that is not a choice any serious leader should make.

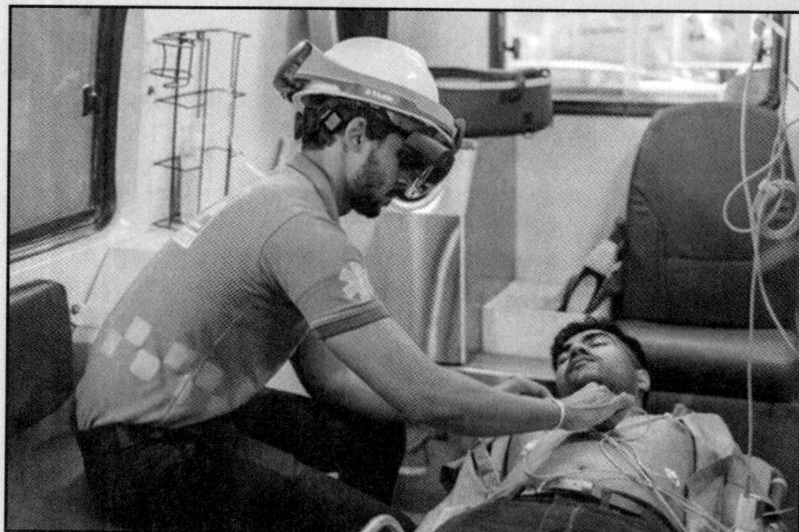

FIGURE 8–2. A Sri Lankan EMT utilizing augmented reality during patient care.

CASE STUDY

Sacramento County's Artificial Intelligence–Powered Ambulance Patient Offload Times Reduction Initiative

In 2023, Sacramento County EMS confronted a challenge that plagues many systems nationwide: excessive ambulance patient offload times (APOT)—for years, average offload times had exceeded 70 minutes, straining EMS availability, eroding clinician morale, and worsening outcomes for patients waiting for hospital beds. Traditional tracking methods—manual timestamps, fragmented handoff records, and reactive reporting—failed to generate real-time insight or actionable solutions. What Sacramento needed was not another policy—it required a system transformation.

Working in collaboration with AppOrchid, a California-based AI platform company, Sacramento County launched a first-of-its-kind pilot project to apply AI and real-time data orchestration to the APOT crisis.[11] Rather than adding staff or rewriting regulations, the project focused on integration: merging EMS data (from platforms like ImageTrend and ESO) with hospital electronic health record data (from Epic and Cerner), creating a unified "data lake" of real-time operational intelligence.

AI algorithms monitored real-time ambulance geolocation, emergency department (ED) registration status, and hospital capacity data. Using geofencing technology, the system automatically recognized ambulance arrivals and began preregistration workflows. Predictive analytics projected when patients would be offloaded based on ED crowding trends and historical throughput rates. Instead of waiting passively for availability, EMS and hospital teams could anticipate bottlenecks and prepare accordingly.

The result? This is not just about faster offload times but also about better alignment between EMS and emergency departments. Early findings showed a measurable reduction in prolonged delays, increased ED situational awareness, and a significant decrease in wall time for EMS crews. Even more important, the platform provided something EMS rarely has: shared visibility. With both EMS and

(*continues*)

hospitals accessing the same live data, mutual accountability replaced finger-pointing, and collaboration replaced isolation.

The Sacramento APOT initiative is more than a local success story. It is a national proof of concept. It shows how AI-powered platforms, when deployed with intentional data integration, real-time logic, and stakeholder engagement, can transform not only how EMS works but also how EMS connects to the more extensive healthcare system.

This case is not about a new tool. It is about a new mindset: real-time decision support is not just a luxury for trauma teams and critical care units. It is a necessity for every EMS system that wants to move from reactive response to proactive coordination.

Clinical, Legal, and Operational Implications

Real-time decision support is moving fast. While still novel in EMS, these systems are already transforming hospital care, particularly radiology. AI now assists with scan interpretation, reduces diagnostic errors, and catches subtle findings sooner.

Now ask yourself: If your family member had a radiology test, and the physician missed the cancer diagnosis, would you be okay knowing AI support was available but *was not* used when it could have made the difference?

Probably not.

This is where medicine is headed. We are approaching a tipping point where the liability is not for using AI but for not using it. When real-time clinical decision tools become the standard of care, *failing to adopt them will carry consequences.*

If the tech exists to flag early sepsis, catch a missed ECG finding, or suggest time-sensitive interventions—and we do not use it—that is exposure. This is not just a clinical risk; eventually, it will increase liability risks.

Human Judgment, Augmented

You were trained to assess, think critically, and act fast. For years, however, we have asked EMS clinicians to make complex decisions under impossible conditions—without support, data, or a safety net.

CASE STUDY

Physicians Using Artificial Intelligence Caught More Cancer—with Fewer False Alarms

From 2021 to 2023, hospitals across Germany studied how AI could support physicians reading mammograms. Over 460,000 women were screened. Some physicians chose to use AI tools—others did not.

- **Physicians who used AI found more cancer.** They detected **nearly 18% more cases** of breast cancer than physicians who did not use AI.
- **AI did not cause unnecessary callbacks.** Despite better detection, the number of patients sent for follow-up was **slightly lower** when AI was used.
- **Accuracy improved.** When the AI-supported team flagged something, they were right more often. The **positive predictive value**—the chance that a flagged finding was cancer—was significantly higher.

When clinicians lead and AI supports them, outcomes get better.[12]

When properly designed and implemented, real-time decision support does not override your experience or clinical judgment—it sharpens your awareness. It amplifies your ability to make the right call before things get worse.

Think about the difference between guessing and knowing. I hope you are right and seeing the pattern confirmed on your augmented reality heads-up display. AI, when properly integrated, restores your bandwidth. It offloads the background noise and highlights the signals that matter most. You still decide—but you decide with better input, earlier insight, and stronger confidence.

This is not about replacing intuition or gut instinct. It is about reinforcing both with machine-speed feedback that aligns with your protocols and enhances your training.

The medic in a dimly lit living room at 2 a.m. still decides. However, now, they are not making that decision alone.

This is the future of EMS leadership: AI behind the scenes, clinicians in charge.

Know What You Are Using

We have discussed the power of AI in clinical documentation, information retrieval, and clinical decision-making, but there is a critical distinction you must understand:

- **Large language models** used for documentation, such as AI-generated narratives or voice-to-SOAP (structured subjective, objective, assessment, and plan format), are *not* regulated by the FDA. They support administrative tasks and efficiency, not medical judgment. The clinician reviewing the report still bears the liability.
- **Clinical decision support models**—tools that suggest diagnoses, recommend interventions, or flag treatment pathways—are medical devices that require FDA approval because they influence patient care decisions.

If your EMS agency is using AI to trigger alerts (sepsis, stroke, STEMI [ST-segment elevation myocardial infarction]), you are already entering clinical decision support models territory.

Key Terms and Definitions

- **Augmented Reality:** Technology that overlays digital information (such as vitals or procedural guidance) onto the physical environment, often through smart glasses or headsets, to support real-time clinical decision-making.
- **Cognitive Load:** The mental effort required to process information and make decisions, often heightened in EMS due to high-acuity situations, limited resources, and multitasking demands.
- **Connected Ambulance:** An EMS vehicle equipped with advanced technology—including AI-enabled devices, real-time telemetry, augmented reality tools, and remote physician access—to enhance clinical support and care coordination during transport.
- **Data Lake:** A centralized digital repository that stores structured and unstructured data from multiple sources, allowing AI platforms to conduct real-time analysis and generate predictive insights.
- **FDA-Regulated Clinical Decision Support Model:** AI systems that provide diagnostic or treatment recommendations. Because they influence clinical decisions, they are considered medical devices subject to FDA oversight.

- **Modified Shock Index (MSI):** A calculated ratio (heart rate divided by systolic blood pressure) used to identify early hemodynamic instability, often integrated into AI decision tools for early detection of clinical decline.
- **Real-Time Decision Support:** AI-assisted platforms that analyze live clinical data and environmental inputs to deliver time-sensitive recommendations, warnings, or prompts during active patient care.
- **Sepsis Prediction Model:** Epic's embedded AI algorithm analyzes patient data continuously to detect early signs of sepsis and prompt clinician response within the electronic health record workflow.

Reflection Questions

1. Where in your system would real-time AI decision support have the most significant impact—stroke alerts, trauma triage, or medical calls? Why?
2. Think back to a case where a subtle change in vitals or condition caught you off guard. How might AI have changed that call's outcome?
3. What challenges would your agency face in adopting real-time AI tools—and who needs to be part of solving them?
4. What policy, training, or oversight changes would you want in place before trusting an AI-generated alert in the field?
5. How can we ensure that real-time AI tools are equitably available in rural, volunteer, and resource-limited EMS systems, not just large urban ones?
6. If failure to use AI becomes a liability issue, what will EMS need to do now to avoid falling behind or exposing risk?

Study Questions

1. According to the chapter, what is the most critical value of real-time decision support tools in EMS?
2. Which of the following is NOT a feature of AI-assisted EMS dispatch systems?

   a. Voice stress detection
   b. Live language translation
   c. Automated ambulance driving
   d. Predictive unit deployment

3. True / False: Real-time AI systems are designed to replace EMS clinicians by automating clinical decisions.
4. What EMS challenge does Sacramento County's AI-powered APOT initiative attempt to solve?
5. How did Sri Lanka's 1990 Suwa Seriya EMS system integrate AI and augmented reality into national operations?
6. Which of the following are legal or ethical concerns related to AI use in EMS?

    a. Algorithmic bias and data privacy risks
    b. Reduced patient satisfaction scores
    c. Increased cost of ambulance maintenance
    d. Decreased use of EMS protocols

7. Real-time AI decision support tools are only as effective as the _______ and oversight systems governing their use.
8. What role can AI-powered translation tools play in improving EMS care?
9. How can MSI trends be used by real-time AI systems to assist EMS clinicians?
10. Which of the following leadership actions is essential for effective AI integration in EMS?

    a. Eliminating all paper protocols
    b. Centralizing dispatch under federal control
    c. Building systems that support clinician oversight and transparency
    d. Mandating AI use without training

CHAPTER SUMMARY

Real-Time or Left Behind

AI is already reshaping real-time clinical decision-making across the global healthcare system. In hospitals, predictive models for sepsis, AI-assisted imaging diagnostics, and continuous monitoring alerts are standard practice, saving lives by recognizing deterioration before it becomes irreversible. Health systems in countries like Sri Lanka are deploying AI-supported triage and care coordination tools at the national level, proving that scalable, real-time intelligence is not reserved for the future—it is operating now.

Meanwhile, EMS remains largely disconnected from this shift. The technology exists, the models are validated, and the outcomes are measurable. However, without focused leadership and coordinated action, EMS risks being excluded from the next generation of real-time decision support infrastructure.

This chapter makes the case for urgency. AI-enhanced dispatch and documentation platforms already demonstrate that clinical quality, legal protection, and operational efficiency can be simultaneously improved. Tools that identify threats buried in background noise, optimize ambulance deployment based on predictive analytics, and guide protocol adherence in real time are not theoretical—they are in use.

The core message is clear: Real-time AI support does not replace EMS clinicians. It strengthens their judgment, increases system visibility, and protects patients when seconds count.

The question is not if AI belongs in EMS. The only question left is how fast we can bring it to the field.

9 Drones, Driverless Tech, and the Future of Emergency Medical Services Mobility

LEARNING OBJECTIVES

After completing this chapter, the reader will be able to

- Describe the operational status of AVs and drones, including their presence in EMS-adjacent sectors such as public transit, logistics, and healthcare delivery.
- Evaluate EMS-specific use cases for autonomous systems, including interfacility transfers, low-acuity transports, rural coverage support, and aerial supply delivery.
- Analyze the workforce implications of autonomous ambulances, particularly how single-clinician models and automated driving can alleviate staffing shortages and reduce burnout.
- Identify outdated regulatory barriers, such as dual-crew requirements or driver's license mandates, that may obstruct innovation, and explain strategies for regulatory modernization.
- Compare the safety performance of human-driven ambulances and AVs, using current crash rate data and risk analyses from organizations like NHTSA and Waymo.
- Apply lessons from the transportation sector (e.g., taxi industry disruption) to predict how nonemergency EMS transport may be vulnerable to autonomous tech-based competitors.
- Formulate a strategic EMS leadership response to autonomous technologies, including establishing readiness teams, piloting use cases, advocating for regulatory change, and engaging public stakeholders.
- Explain the operational advantages of drones in EMS, including automated external defibrillator delivery, blood drops, and scene surveillance, and how these capabilities may transform traditional response models.
- Articulate the risks of inaction, particularly regarding loss of market share, regulatory voice, and public trust if EMS does not lead in integrating emerging transport technologies.

Leadership Perspective: This Is Not Science Fiction—It Is Logistics

In early 2024, I recorded a social media interview about the future of autonomous vehicles (AVs) in EMS. I talked about the rise of driverless technology, noting that AVs are no longer experimental. They are operating today in major U.S. cities. This point caught people off guard: the transition from robotaxis to medical transport is not a technology shift—it's an operational shift. Right now, the same system Waymo installs on a Jaguar can be mounted on an ambulance.

That clip received thousands of views and tens of thousands of comments. Many were shocked, not because the idea was far-fetched but because they had never considered the existing technology. Here is the irony: While the EMS community debated hypotheticals, the private sector quietly prepared to move forward.

Driving is arguably the most dangerous part of the job for emergency medical technicians (EMTs) and paramedics. Ambulance crashes occur regularly—often weekly—and are rarely caused by mechanical failure or system malfunction. They happen because a human made a mistake. Human error is a factor in 94%[1] of all fatal vehicle crashes. The risk is real and persistent, whether the at-fault driver is behind the ambulance wheel or in the oncoming lane.

AVs have the potential to significantly reduce that risk. Waymo alone has logged millions of miles with dramatically fewer collisions and substantially lower injury severity than human drivers. For EMS, the opportunity is not just about automation—it is about safety, consistency, and protecting the clinicians and patients in transit.

One of the major objections I hear is, "But EMS requires lights, sirens, red-light runs, wrong-way driving, etc." And my answer: Maybe it should not. The data is clear—lights and sirens provide minimal to no benefit in most EMS calls. What they *do* increases the risk to both patients and the public.

Now, I am not saying autonomous ambulances will replace *all* EMS operations. They will not. Not anytime soon. However, could they augment what we do today? Could they help solve the staffing crisis? Absolutely.

Think about rural and frontier communities. Many rely on volunteers. When a call comes in, they drive to the station and the scene. But what if an autonomous ambulance was dispatched directly to the patient, and the volunteer clinician met it there? What if interfacility transports did not require full advanced life support (ALS) crews sitting idle for three hours on the highway? What if a single medic could be in the back providing care without needing to drive?

We must recognize that most EMS patients are transported with one provider in the back. The second provider is, by necessity, driving. If autonomous technology can remove that burden, we can redeploy talent, reduce costs, and maintain services in areas on the brink of closure.

Let's not miss the bigger picture. The most expensive part of EMS is not equipment or fuel—it's people. We're already short-staffed; agencies are closing shifts, stations are going dark, and overtime is unsustainable. If autonomous ambulances can rebalance workload, reduce risk, and help us retain access in underserved areas, we should be first in line to explore them.

This is not science fiction; it is a strategy. Now, let's zoom out. Before talking about autonomous ambulances, we must understand the landscape of AVs.

Autonomous Vehicles: The Road Is Already Paved

Before discussing autonomous ambulances, we must understand the broader landscape of AVs. This technology may still feel theoretical for many EMS clinicians and leaders, but it is not. It is active, expanding, and transporting hundreds of thousands of Americans weekly. What follows are the operational benchmarks at the time of this writing and by the time you read this, the numbers will be larger.

Deployment Across the United States

AVs are operating in multiple U.S. cities with state approval. Waymo, a leader in this space, has launched commercial robotaxi services in Phoenix, San Francisco, Los Angeles, and Austin. By the middle of 2024, Waymo had completed over 100,000 paid autonomous rides per week across nearly 500 square miles of service area. That is not a trial—that is scale.

Other companies, including Cruise Automation (a division of General Motors), have operated in Phoenix, Dallas, and Houston. Although Cruise faced regulatory setbacks, the trajectory of AV deployment remains forward-leaning. More than half of the United States now allow some form of AV use or testing under defined conditions.

These vehicles are on the road, not in research labs.

Safety and Transparency

Public trust remains a hurdle, but the safety data is compelling. From 2019 to mid-2024, AVs were involved in nearly 4,000 reported incidents, including 473 in 2024. Many of these were low-severity, and the most common cities for AV collisions—San Francisco, Phoenix, and Los Angeles—are also where AVs are deployed at scale.

Waymo, in particular, stands out for its transparency. Over 100 million rider-only miles have been logged to date. According to their 2024 data hub, when compared to human drivers in similar environments, Waymo vehicles have

- 84% fewer crashes involving airbag deployment,
- 73% fewer injury-causing crashes, and
- 48% fewer police-reported crashes.

Waymo's crash rate analysis is peer-reviewed and follows the RAVE (Retrospective Automated Vehicle Evaluation) safety framework, giving communities, policymakers, and researchers a replicable, evidence-based way to assess performance.[2]

Public Adoption and Use Cases

Still, many Americans remain cautious. As of 2022, only 37% of the public said they were willing to ride in a vehicle that is fully autonomous. That number is slowly rising as riders report overwhelmingly positive experiences—more space, less stress, and zero risk of distracted or intoxicated driving.

AVs are used for short urban trips, airport transfers, and neighborhood-level transit. In 2024, the most popular destinations included Phoenix Sky Harbor Airport, The Ferry Building in San Francisco, and The Grove shopping center in Los Angeles.

Waymo even reported avoiding over 18 million kilograms of CO_2 emissions in 2025 by deploying an all-electric, renewable-powered fleet.

Why Emergency Medical Services Should Pay Attention

This is not just a transportation shift but a public safety shift. AVs offer new pathways to move people, resources, and clinicians more safely and efficiently. Moreover, the most crucial part is that they are not coming—*they are already here.*

As we transition into the EMS-specific implications, it is critical to understand that this is not a question of *if.* It is a question of *how*—and whether EMS will lead that integration or be displaced by those who do.

From Concept to Contact: The Rise of Autonomous Emergency Medical Services

While AVs are gaining traction in the consumer world, EMS-specific applications are already beginning to take shape. The transition is underway from urban shuttles to conceptual ambulances—not in theory but in practice. Like other sectors, EMS must decide whether to lead or react to the change after the fact.

Volkswagen's ID. Buzz ambulance concept—built on the company's modular electric drive platform—has also captured international attention. [3] It's not just a showpiece. Road-testing of the autonomous variant began in 2023, and by May 2024, Volkswagen announced plans to bring the non-autonomous version of the ID. Buzz to the U.S. market. While a launch date for the fully autonomous ambulance version has not yet been confirmed, the convergence of consumer robotaxi deployment, commercial autonomous trucking, and advanced air mobility aerial pilots means the technical and regulatory pieces for a driverless ambulance are now assembling in parallel, faster than any single sector could deliver alone.

However, EMS does not need to wait for a prototype. The Beep ROBO Ride Medical pilot program in Peoria, Arizona demonstrated that AVs can operate reliably within a complex healthcare corridor. Partnering with the Maricopa Association of Governments, Beep launched a fully electric, driverless shuttle system to connect over 100 medical facilities and more than 500 senior living residences.[4] This is not a concept. It is a functioning component of Peoria's healthcare infrastructure (fig. 9–1).

Innovation is also emerging from the academic world. At Cedarville University in Ohio, engineering students have been building and testing autonomous ambulance designs with life support systems and real-world interface features.[5] These are not fantasy models. They are functional systems that integrate with today's clinical and dispatch technologies.

So, what is the opportunity for EMS?

AVs will not replace traditional 9-1-1 ambulances tomorrow. But they can—and should—be integrated into EMS systems very soon in specific, high-value use cases:

- **Interfacility transfers** that tie up advanced life support crews for hours
- **Scheduled transports** for nonurgent care

FIGURE 9–1. ROBO Ride Medical was a pilot program in Peoria, Arizona, operated by Beep, Inc., in partnership with the City of Peoria and the Maricopa Association of Governments. The project provided free, autonomous shuttle rides for residents traveling to nearby medical facilities, using electric, driverless vehicles monitored by onboard attendants. Conducted in collaboration with Arizona State University, the initiative explored the feasibility and public acceptance of AVs for healthcare transport. The project demonstrated the potential utility in future EMS and patient mobility systems.

- **Low-acuity calls** that still require a physical response but not full-staffed transport
- **Rural coverage zones** where volunteer or part-time staffing makes conventional deployment unsustainable

In these situations, an autonomous ambulance could be dispatched with a single EMT or paramedic in the patient compartment or meet a clinician at the scene. In either case, the driver becomes the caregiver. The unit becomes a mobile care pod, not just a vehicle.

It is tempting to think of these innovations as far off. But that mindset is exactly what doomed the taxi industry. Uber and Lyft did not blindside traditional cab companies. They saw it coming; they just failed to act. Now EMS cannot afford to make the same mistake. The interfacility and scheduled transport sectors—often critical sources of revenue for EMS agencies—are particularly vulnerable to disruption. And once they're lost, they will not be coming back.

The public narrative around AVs is rapidly shifting. Cities like San Francisco and Phoenix are normalizing the presence of driverless vehicles in everyday life. As public familiarity grows, the resistance to seeing an ambulance without a front-seat driver will decline. EMS leaders must not ask whether autonomous ambulances are viable but whether we are prepared to use them strategically before someone else uses them against us.

The opportunity is here. The question is whether EMS will embrace it early and shape the system, or arrive late and inherit whatever model others build.

The Workforce Equation

If AVs offer one thing EMS needs more than anything right now, it is relief.

Across the country, agencies grapple with the same crisis: clinician shortages, forced overtime, and unsustainable staffing patterns. Coverage gaps have become routine, and entire shifts go unfilled. Volunteers are aging out in rural areas, and replacements are not moving into the metro systems. Burnout and attrition hollow out even the best-equipped departments.

Let's be blunt: Our current model is strained and breaking.

AVs will not solve the workforce crisis overnight, but they may finally allow us to use the workforce we have more effectively. Consider the structure of a typical EMS response: one paramedic in the back and one EMT driving. For most transports, the provider behind the wheel contributes no patient care because they cannot. That clinician is locked into the least productive role in the system.

Now, imagine that the clinician redeployed. Sometimes, it's a second set of hands in the patient compartment. In others, it's an ALS provider assigned to another call. Or it's one less overtime shift that burns out your full-time staff. That is what autonomous transport can unlock—not replacement, but reallocation.

This is especially relevant for systems that already operate on thin staffing margins. Picture a rural community where only one EMT is on duty. If an autonomous ambulance can pick them up and deliver them to the patient, or rendezvous on-scene, that unit becomes viable again. Volunteers no longer need to drive to a station, start a rig, and burn minutes just getting moving. The vehicle moves first. The care arrives faster.

Urban systems have a different challenge: unit hours lost to low-acuity transports. ALS units tied up for hours on routine interfacility transfers take resources away from 9-1-1 calls. What if AVs handled those scheduled transports instead? What if those crews were freed up for time-critical emergencies?

The benefits are not just clinical—they're financial. Staffing is the most expensive part of any EMS operation. Personnel costs dominate every budget, even in systems where pay is below market value. Labor shortages drive higher costs and force agencies into unsustainable overtime, hiring bonuses, and contract labor.

Autonomous transport offers a new lever to pull.

This is not about reducing the workforce, but for some EMS agencies, this technology could be the difference between remaining operational and closing. If properly implemented, this may assist with preserving access, maintaining coverage, and ensuring that EMS clinicians can focus on care, not the cognitive load of navigating traffic or the burden of back-to-back transports without rest.

Most EMS patients do not need two clinicians. Most calls do not require advanced driving skills. Most transports do not need to burn through a whole crew to complete. And if we can staff smarter, we can stretch farther.

Autonomous ambulances will not solve the workforce crisis. However, they will give us a tool we do not have today—and a fighting chance to keep our systems intact.

Lessons from the Taxis: A Warning Shot

When Uber and Lyft entered the market, taxi companies did not panic. They scoffed. They pointed to their city contracts, union protections, and long-standing brand recognition. They assumed that being established was enough. They believed the public would stay loyal. They were wrong.

The taxi industry did not collapse because it lacked infrastructure, but because it lacked vision.

Trips per Day in New York City: Taxi versus Ride Share

That same blind spot now threatens EMS. At the same time, we debate whether autonomous ambulances "fit" within our current service models. AV technologies are advancing—not just in concept but in competitive application.

In the private sector, robotaxi companies are now directly challenging rideshare disruptors like Uber and Lyft (fig. 9–2). In the logistics space, AV developers are targeting hub-to-hub trucking operations—routes between transfer hubs or distribution centers—because they are predictable, repeatable, and economically essential. The parallels to EMS interfacility transport are striking.

EMS agencies across the country complete scheduled trips between Hospital A and Hospital B multiple times daily. Dialysis transfers, behavioral health transports, and interfacility medical moves are not exceptions—they are

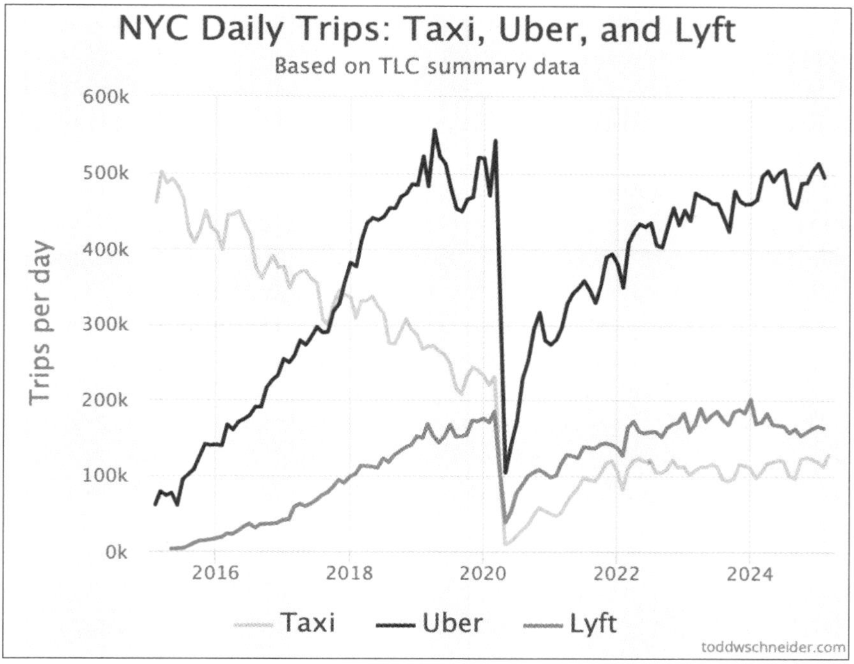

FIGURE 9–2. Data from New York City's Taxi and Limousine Commission shows the dramatic rise of rideshare services since 2015. Taxi ridership collapsed as Uber and Lyft surged, serving as a cautionary economic tale for EMS leaders facing disruptive mobility technologies.[25]

foundational. For many systems, these calls help subsidize 9-1-1 response, justify vehicle acquisition, and maintain a stable workforce.

And they are ripe for disruption.

If EMS does not proactively shape how autonomous transport enters health care, these high-frequency, high-efficiency trips will be captured by third-party vendors, not because they offer better care, but because they provide more convenience, reliability, and scale. Once those transports leave our ecosystem, they will not return. We risk losing not just the revenue, but also the workforce leverage, regulatory voice, and ability to ensure systemwide readiness.

The taxi industry's lesson is instructive: Disruption did not come from better cars—it came from better systems. Patients and facilities were not won over by cost savings but by convenience, control, and transparency.

Autonomous transport offers the same wedge.

Facilities do not need to be convinced that an AV ambulance is cheaper. They only need to believe it is faster, easier to schedule, and guaranteed to arrive on time. And once that becomes true, why call the local EMS agency?

This is not theoretical. It is a visible trajectory. McKinsey forecasts an over $600 billion global market for autonomous trucking by 2035, driven by predictable routes and rising labor costs.[6] Their model begins with hub-to-hub deployment—not unlike the scheduled interfacility trips EMS provides every day (fig. 9–3).

The threat is not that someone will build a better ambulance. The danger is that someone will create a more innovative logistics system that separates transport from care and convenience from clinical judgment. If that happens, EMS will be left reacting to a reality we did not help define.

For many EMS agencies, losing interfacility transports would be catastrophic. Without them, the financial ability to cross-subsidize high-risk, low-revenue 9-1-1 calls would disappear, and with it, so would readiness.

If EMS wants to stay relevant in a world of autonomous medical transport, we must lead—not lag—the development of operational models, data standards, staffing policies, and safety protocols. The time to act is not when disruption arrives; it is when it begins to take shape (fig. 9–4).

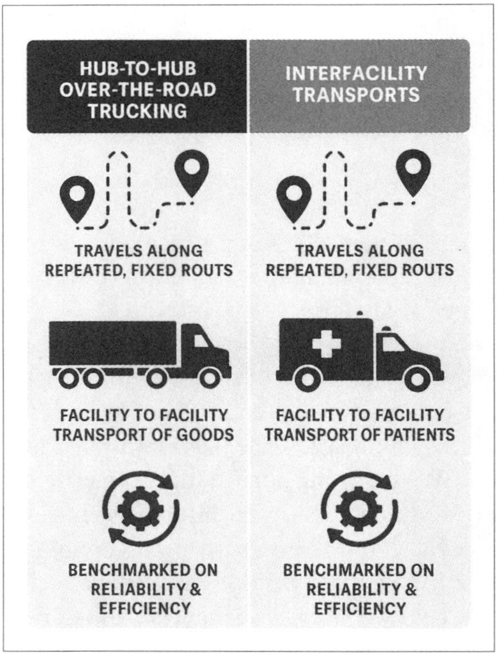

FIGURE 9–3. The similarities between freight logistics and interfacility EMS transports reveal a clear pathway for automating these processes. Both rely on predictable, fixed routes where reliability, safety, and efficiency can be benchmarked and improved.

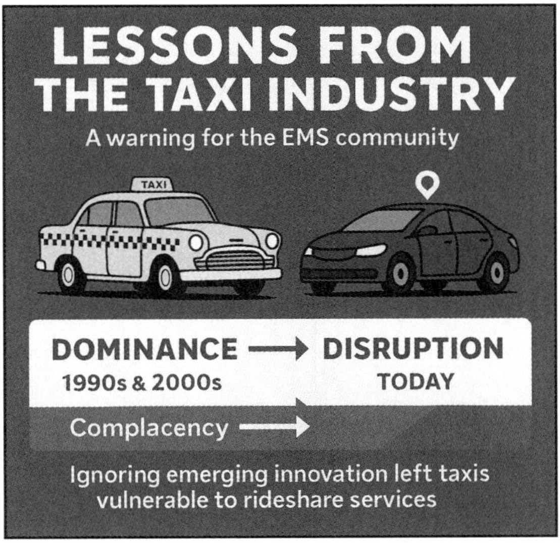

FIGURE 9–4. Like taxis in the early 2000s, EMS risks disruption if it ignores technological innovation. Complacency toward automation, AI, and evolving mobility models could leave traditional EMS systems unprepared for the next wave of change.

Drones in Emergency Medical Services

While autonomous ground vehicles often dominate the conversation, aerial autonomy may prove even more disruptive. And in some places, it already has.

In Rwanda, drones are not experimental—they are standard operating procedure. Zipline, a fixed-wing drone delivery company, has completed hundreds of thousands of blood deliveries to rural clinics nationwide. These drones fly in all weather, operate autonomously, and drop lifesaving units of blood, vaccines, and medications with parachute accuracy. The result: stockouts have decreased, delays are rare, and patient outcomes have improved.[7]

That same model is being explored here at home.

The New York City Fire Department[8] has begun testing drone-based delivery systems that deploy blood and trauma supplies to field medics. These units can navigate dense urban terrain, avoid traffic delays, and deliver critical resources where responders need them most. Whether flying to a high-rise rooftop or touching down at the edge of a park, drones eliminate the delay between arrival and access.

The U.S. military is doing even more. In combat zones and remote outposts, autonomous aerial systems now deliver blood, resupply kits, and advanced medications directly to medics. These systems operate beyond visual line of sight, fly autonomously, and are programmed with mission-specific logic. What began as a battlefield innovation is now entering the civilian EMS environment.

So what does this mean for us?

It means a new tier of response is emerging—one where not every mission requires a whole crew, a staffed rig, or even a physical presence. Here is what drones could do for EMS:

- **Automated external defibrillator and naloxone delivery:** In rural areas, a drone could arrive minutes ahead of the ambulance, delivering lifesaving interventions before hands-on care is possible.
- **Blood and medication drops:** In wilderness rescue or long extrication scenarios, drones can deliver high-value assets without tying up air or ground units.
- **Mass casualty triage:** Aerial systems equipped with cameras, thermal imaging, or even loudspeakers can begin triage, communicate with victims, and transmit live intel to incoming responders.
- **Telemedicine and scene surveillance:** With 5G connectivity, drones can serve as flying cameras or communication relays, enabling physicians or command staff to assess conditions in real-time.

Rural afnd volunteer systems, meaning the communities most strained by workforce attrition and geographic distance, may see the most significant near-term benefit. In Hyde County, North Carolina, volunteer paramedics are participating in a proof-of-concept program using Pivotal's electric ultralight Electronic Vertical Takeoff and Landing (eVTOL) aircraft, which qualifies under FAA Part 103 rules and requires no pilot certification—only a two-week in-house training course.[9] The program's initial purpose is not patient transport but rapid transport of EMTs and paramedics to the scene ahead of a ground ambulance, with the aircraft remaining able to bypass traffic and terrain that can slow conventional response. This is exactly the kind of low-infrastructure, low-cost, high-impact model that a rural volunteer system might actually be able to uphold, and it is the first of what will likely be many such programs built on the eVTOL Integration Pilot Program precedent.

The cost of these systems is dropping, the capabilities are growing, and the regulatory framework is no longer catching up. It is sprinting. Under Executive Order 14307,[10] the FAA established the eVTOL Integration Pilot Program (eIPP)[11] and announced the first awardees in March 2026, with initial operations required within 90 days of selection. One of those awardees is the North

Carolina Department of Transportation,[12] whose Electric Logistics and Integrated Flight Testing-NC program partners with Beta Technologies and Joby Aviation to build a statewide network of electric aircraft connecting rural hospitals, airports, and communities, explicitly focused on improving access to health care. A separate North Carolina effort in Hyde County pairs volunteer paramedics with Pivotal's Part 103 ultralight eVTOL as a proof of concept for accelerating rapid response in rural areas.[13] Metro Aviation, one of the nation's largest air ambulance operators, has placed a deposit-backed order for up to twenty Beta Alia eVTOL aircraft to be integrated into its existing fleet,[14] and the U.S. Department of Health and Human Services has awarded Beta a research contract to study Alia's emergency response capabilities.[15] At the same time, Sikorsky delivered to the U.S. Army the UH-60MX Black Hawk,[16] the first full-authority fly-by-wire, optionally piloted Black Hawk—fully integrated with the MATRIX autonomy suite—and unveiled the autonomous R66 TURBINETRUCK with Robinson Helicopter.[17] MATRIX has now been validated across 21 aircraft platforms with more than 1,000 flight hours of operational data. Autonomous and electric aviation are no longer speculative; they are fielded technology with named EMS agencies, aircraft, and programs. Whether an EMS agency directly participates in the eVTOL Integration Pilot Program or not, these pilots will generate the operational playbooks and regulatory precedents that define EMS access to aerial medical delivery for the next decade. EMS leaders who are not already exploring partnerships in this space are not merely behind the curve; they are becoming the obstacle that others are bending around.

Let's be honest—many still think of a toy when we picture a drone today. However, the systems being deployed by public safety agencies and health ministries are anything but toys. They are strategic assets that are changing the geography of care.

The choice EMS faces is not whether drones belong in our system, but whether we'll either use them or watch someone else fly past us (fig. 9–5).

Regulation, Risk, and Relevance

Technology moves fast. Regulation rarely does.

If autonomous ambulances are to play a meaningful role in the EMS landscape, we must confront a hard truth: Our laws and rules were not written for this. Much of what governs EMS operations—especially at the state level—is rooted in tradition, not evidence. And tradition is a poor guide when you are building the future.

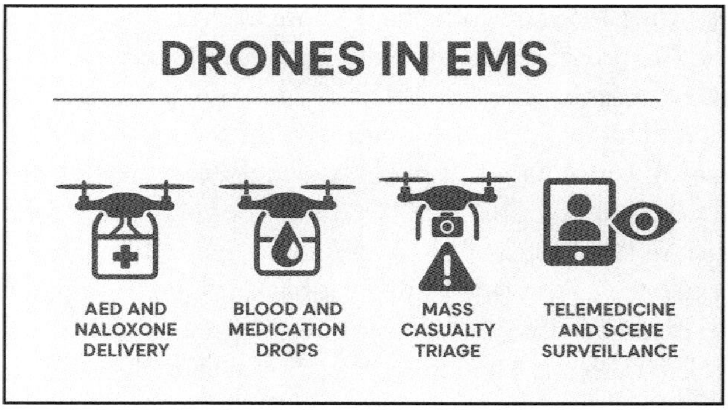

FIGURE 9–5. Drones extend EMS capabilities beyond traditional response. From naloxone delivery to mass casualty triage, these platforms provide rapid logistics, situational awareness, and telemedicine access in hard-to-reach areas.

Across the country, state EMS rules still reflect outdated assumptions. In some states, you cannot be licensed as an EMT or paramedic unless you hold a valid driver's license. That requirement makes sense—if we assume that every EMS clinician must also be a vehicle operator. But what happens when they're not? What happens when the "driver" is autonomous and the clinician only cares for the patient?

The authority to issue or revoke a driver's license belongs to the Department of Motor Vehicles, not the EMS regulatory body. The choice to hire an EMS clinician with or without a driver's license should be the employer's preference. Linking the two may have made historical sense, but it creates unnecessary barriers today. It is time to reexamine whether driving ability should affect clinical licensure.

There are other structural issues just like it.

Many states still require two certified EMS clinicians in every transport unit. While this is standard practice, it is rarely evidence-based. For most EMS transports, one provider is in the back, and the other is simply driving—no patient contact, no intervention, no clinical decision-making, just miles on the road.

When autonomous ambulances enter the fleet, we must ask hard questions: Do we need two clinicians for every patient? Or has the regulation outlived its relevance?

Some will raise safety concerns. What if a patient becomes violent? What if lifting is required? What if something goes wrong?

These are all valid questions—but none without answers: triage and patient selection matter. Not every patient is a candidate for single-provider autonomous transport. However, a stable patient with a minor injury, a routine interfacility transfer, or a behavioral health transport with remote monitoring support may all be viable use cases.

Others will point to workforce standards. "We are not ambulance drivers," they say—and they are right. However, ironically, some of our policies treat us exactly like that. When one EMS clinician is relegated to driving alone, with no clinical input, that role becomes indistinguishable from a nonmedical transport function. If we want to elevate EMS, we must also be willing to redesign it.

This conversation will not be easy, but it is essential.

State EMS offices, medical directors, legislators, and agency leaders must begin evaluating which regulations are still necessary and which are barriers to innovation. These discussions must include EMS clinicians and also the communities we serve. Transparency and inclusive stakeholder engagement are nonnegotiable. We cannot build trust with the public unless we bring them to the table.

Change must be deliberate, and change must also be courageous.

The regulatory landscape can be a launchpad or a wall. EMS will determine which one we'll build.

Thankfully, EMS does not have to start from scratch. The U.S. Department of Transportation—through the National Highway Traffic Safety Administration's (NHTSA's) *Automated Driving Systems: A Vision for Safety 2.0*—has already outlined a flexible, risk-based framework for deploying AV systems safely and ethically. Their guidance emphasizes transparency, human oversight, and system-level accountability, aligning with EMS values. NHTSA calls on public safety stakeholders—including emergency services—to help define the operational design domains, human-machine interfaces, and data standards that will shape autonomous systems in high-risk environments. For EMS, this is not just an opportunity but an obligation to lead the policy conversation, not wait for decisions made without us.

Rethinking Risk: Safety at the Core

If AVs are to have a legitimate role in EMS, they must demonstrate one thing above all else: They must be safer than what we have now. Fortunately, the evidence is beginning to point in that direction—and the contrast with our current reality is stark.

Let's start with what we know. Ambulance crashes are a leading cause of death for EMS clinicians in the line of duty.[18] According to the 2023 *Analysis of Ground Ambulance Crash Data* by NHTSA, there was an average of 24.7 fatal crashes and 28.4 fatalities per year from 2012 to 2018.[19] Of those killed, 40.2% were ambulance occupants, including drivers, crew members, and patients. The remainder were primarily occupants of other vehicles, with a smaller percentage involving pedestrians or cyclists.

This is not just a statistical issue. Behind each number is a broken body, a grieving family, a lawsuit, or a shattered community. More than 92% of special crash investigations reviewed by NHTSA involved some operator error, including running red lights, poor intersection clearing, or fatigue. Even more concerning, only 8.8% of clinicians in the patient compartment were properly restrained during crashes. In many agencies, the term ambulance driver remains a cultural joke—until it is not.

Now, contrast that with the data from Waymo in Phoenix, San Francisco, and Los Angeles. In its 2024 publicly validated report, the Waymo Driver logged over 7.1 million rider-only miles—that is, miles driven with no human at the wheel—and demonstrated crash rates dramatically lower than human benchmarks:[20]

- 85% reduction in crashes involving injuries (0.41 per million miles vs. 2.78 per million for humans)
- 57% reduction in police-reported crashes (2.1 per million miles vs. 4.85 per million for humans)[21]

That translates into 17 fewer injuries and 20 fewer police-reported crashes than expected had human drivers been behind the wheel for the same mileage.

Waymo's results are not just statistical outliers but the outcome of a rigorously evaluated, peer-reviewed methodology. The Insurance Institute for Highway Safety, the University of Michigan Transportation Research Institute, and other independent experts have praised the data as "the strongest evidence yet that AVs have lower crash rates than humans."[22]

Now, let's put this in perspective.

Ambulance crashes are not just rare—they are disproportionately more likely to be fatal than similar-sized vehicle crashes. As noted in the overview of the U.S. Fatality Analysis Reporting System data, ambulances crash at a rate of 3,200 per 100 million miles, eight times higher than cars and light trucks. Other sources suggest the actual number may exceed 5,000 per 100 million miles, with crashes being 27% more likely to be fatal than those involving police and 33% more likely than those involving fire vehicles (fig. 9–6).[23]

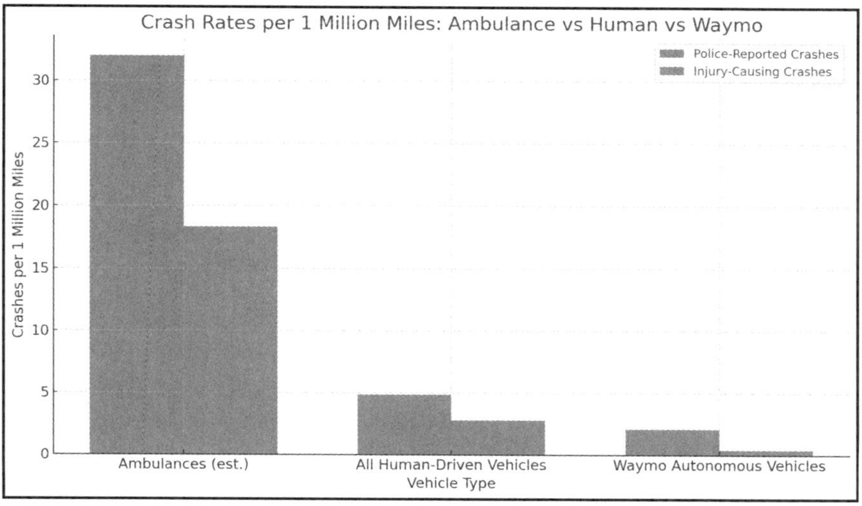

FIGURE 9–6. Ambulance crash rates remain significantly higher than the national average for vehicle crashes. AVs, such as Waymo, demonstrate dramatically lower crash rates, highlighting both the safety potential and policy questions surrounding future EMS automation. Ambulances are estimated to experience more than 6 times the crash rate of all human-driven vehicles and over 15 times the injury crash rate of autonomous Waymo vehicles.[26]

So here is the bottom line:

> *A properly deployed autonomous ambulance, particularly for lower-acuity, prescreened calls, can potentially reduce risk dramatically.*

That does not mean that all EMS transport should be crewless, but it does mean we can no longer afford to ignore safety potential of AVs because of cultural resistance or nostalgic tradition.

We owe our colleagues—and our patients—better.

Strategic Action for Emergency Medical Services Leaders

By now, the question should be straightforward: What will EMS do with this information?

AVs and drones are no longer theoretical. They're here, operating in our cities, reducing crashes, filling transport gaps, and being actively pursued by innovators with no EMS experience, clinical oversight, and accountability to public health outcomes.

If EMS leaders do not act now, others will define this space for us. This is not the time for wait-and-see thinking. This is the moment for strategic movement. Here's what that looks like at the agency, regional, and state levels:

1. **Launch a Technology Readiness Task Force**
   Establish an internal team focused on evaluating autonomous and unmanned systems. The team should include clinical staff, fleet managers, medical direction, dispatch, and regulatory compliance personnel. This is not just about gadgets; it is about clinical workflow, licensing, protocols, and public trust.

2. **Forge Local Partnerships**

   Start conversations with these organizations:

   - City and county transportation departments
   - Academic institutions working on AV or drone tech
   - Public health agencies and hospital systems
   - Regional EMS councils and mutual aid partners

   You do not need to launch a program tomorrow, but you need to be in the rooms where those programs are being shaped.

3. **Identify Low-Risk, High-Value Use Cases**

   Look for places where autonomous tech can be tested without disrupting core 9-1-1 operations:

   - Interfacility transfers
   - Scheduled dialysis or wound care runs
   - Behavioral health transports
   - Rural backup coverage for volunteer systems

   Start small. Start smart. And collect data.

4. **Advocate for Regulatory Modernization**

   Work with your State EMS Office and medical director to identify outdated rules that could block innovation:

   - Does your state require a driver's license to be an EMT?
   - Is two-crew staffing legally required?

- Do EMS agencies have the authority to operate nontraditional or autonomous units?

Be the voice that says EMS clinicians and educators should champion evidence-based regulation over outdated tradition, ensuring that innovation reflects real-world care and protects patient trust.

5. **Engage the Public Early**
   Public trust will make or break autonomous integration. Involve community stakeholders, patient advocates, disability rights groups, and senior service organizations. If your agency has a public education team or public information officer, now is the time to expand their scope to cover emerging technology and safety communication.

6. **Secure External Funding**

   Innovation does not have to break your budget. Pursue these methods of funding:

   - Local transportation or innovation grants
   - State pilot funding (primarily through the Department of Transportation or economic development agencies)
   - Federal grants and pilot participation (e.g., NHTSA, Department of Homeland Security, Department of Health and Human Services, and the FAA's Electric Vertical Takeoff and Landing Integration Pilot Program, including Part 108 beyond-visual-line-of-sight regulations, several of which have explicitly named emergency medical response as a target use case)
   - University partnerships where research grants can cover testing and evaluation

   A $100,000 drone program or AV lease pilot may return tenfold in system stability and public trust.

7. **Report, Share, and Lead**

   Document your efforts, share what works and what does not, submit case studies, present at conferences, and publish findings. This is how EMS moves from isolated innovation to systemwide transformation.

Autonomous technology does not threaten EMS; it tests whether EMS is ready to lead.

Own the Future or Be Outsourced

Here is the question EMS must ask itself: Why are the biggest tech companies targeting transportation?

Why did Uber and Lyft target taxis? Why is Waymo entering a market already served by ridesharing? Why is Amazon investing billions in autonomous trucking and last-mile delivery?

The answer is simple: There is money in motion.

Autonomous systems allow companies to do three things better than traditional models:

- Add value (faster or more consistent service)
- Reduce costs (by removing inefficiencies)
- Increase margin (by scaling without human overhead)

Now, let's turn that lens inward. EMS is a multibillion-dollar market—$18.6 billion in 2023 and projected to grow to over $51 billion by 2033.[24] That is a 10.68% annual growth rate, driven by an aging population, rising chronic disease, and the growing demand for emergency and nonemergency medical transportation (fig. 9–7).

But here is the catch: Most of that growth is not in emergency calls. It's in nonemergency transport, interfacility transfers, dialysis runs, and routine

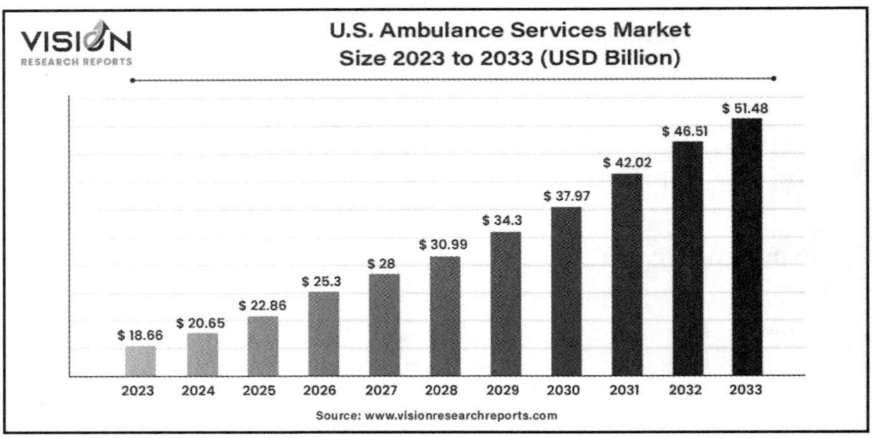

FIGURE 9–7. The ambulance services market is projected to grow from $18.6 billion in 2023 to more than $51 billion by 2033, driven by population aging, healthcare integration, and the adoption of new technologies.

medical appointments. These are the transports that keep many EMS agencies financially afloat. These are the calls that cross-subsidize 9-1-1 services.

Now ask yourself: What happens if a private tech company automates all of that?

What happens if someone builds a reliable, low-cost autonomous system for scheduled basic life support transports, and takes 80% of the calls that are not time-critical, are not trauma, and do not require two clinicians?

Some EMS clinicians might cheer for fewer low-priority transports. But that shift is catastrophic if you run an EMS agency, especially in the private or hybrid sector. Those calls pay the bills. They support the overhead. They allow you to keep units on standby for the calls that *do not* generate revenue.

Strip those transports away, and what is left? A shrinking pool of high-acuity calls, covered by an overworked, underfunded, unsustainable workforce model.

Make no mistake—this industry is ripe for disruption. We're using outdated computer-aided dispatch systems, inconsistent documentation platforms, slow paper-based processes, and staffing models designed decades ago. Meanwhile, the public expects fast, reliable, high-quality service 24/7.

That gap between expectation and capacity is a vacuum. Nature abhors vacuums, and so do venture capitalists.

Now, I do not think traditional EMS will vanish. In the next five years, we will likely not see robotic ambulances with no clinical oversight responding to cardiac arrests. However, I do believe a technological disruption is coming. And I think someone—whether it is Amazon, Waymo, or a startup we have never heard of—will try to capture part of our market. Because they see the same thing, we should see: a fragmented, expensive system with no clear owner and no unified strategy.

We have a choice. We can embrace innovation, partner with technology developers, and build EMS-led models of autonomous integration that preserve clinical oversight, community trust, and financial sustainability.

Alternatively, we can become the next taxi industry: protective, reactive, and irrelevant.

I hope we choose to lead.

Key Terms and Definitions

- **Ambulance Patient Offload Time:** The time between EMS arrival at the hospital and the transfer of care to emergency department staff. A high offload time can lead to EMS unit delays and system strain.
- **Autonomous Ambulance:** A self-driving vehicle equipped for medical transport that can operate without a human driver. In EMS, this vehicle

may support single-clinician configurations or meet clinicians at a scene.

- **Autonomous Vehicle (AV):** A motor vehicle capable of navigating and operating without human input, using sensors, GPS, cameras, and AI-powered systems.
- **Beep ROBO Ride Medical:** An autonomous shuttle pilot program in Peoria, Arizona, that connects medical facilities and senior residences using electric, driverless vehicles.
- **Beyond Visual Line of Sight:** A regulatory classification for drone operations conducted beyond the direct line of sight of the operator, increasingly authorized for public safety and healthcare use.
- **Connected Ambulance:** A transport unit equipped with live telemetry, telemedicine integration, augmented reality, and real-time AI tools to support remote clinician support and decision-making.
- **Driverless Technology:** The hardware and software systems enabling vehicles to operate autonomously, including lidar, radar, cameras, and AI navigation platforms.
- **Drone :** An autonomous or remotely piloted aircraft used for logistics, surveillance, communication, or rapid medical supply delivery in EMS and public safety applications.
- **Interfacility Transport:** Scheduled or nonemergency medical transport of patients between healthcare facilities. This sector represents a significant financial lifeline for many EMS agencies and is vulnerable to autonomous disruption.
- **Technology Readiness Task Force:** A multidisciplinary group within an EMS agency formed to assess, test, and prepare for integrating emerging technologies such as AVs and drones.

Reflection Questions

1. Where in your EMS system could AVs or drones be piloted today, without compromising clinical quality or public trust? Consider scheduled transports, rural backup, or supply delivery.
2. What outdated rules or assumptions are baked into your current staffing, licensing, or operational model? How would these need to change to accommodate autonomous systems?
3. Who in your state or region is already working on AV policy, and are EMS voices represented in those conversations? If not, what would it take to sit at the table?

4. How much of your agency's financial sustainability depends on nonemergency transport revenue? What happens if that business model is disrupted by external competition?
5. Is your agency prepared to lead innovation, or is it waiting for permission? What small, tangible step could you take in the next 90 days to move this forward?

Study Questions

1. Why are AVs particularly relevant to solving the current EMS workforce crisis?
2. Which of the following is **NOT** listed as a current or potential EMS-specific use case for AVs?

   a. Routine interfacility transfers
   b. High-acuity cardiac arrest responses
   c. Low-acuity scheduled transports
   d. Rural coverage where full-time staffing is limited

3. True / False: Waymo's AV crash rates are higher than those of human drivers in urban environments.
4. What lesson does the taxi industry's failure to respond to Uber and Lyft offer to EMS leaders considering autonomous transport?
5. Ambulance crash rates are estimated to be more than times the crash rate of human-driven vehicles.
6. According to the chapter, which two technologies enhance prehospital care delivery in Sri Lanka?

   a. AI-assisted ultrasound and exoskeleton lift devices
   b. Smart glasses with augmented reality overlays and drone-based telemetry
   c. Satellite dispatch and fully autonomous intensive care unit pods
   d. Predictive modeling and robotic triage teams

7. What are two reasons why EMS interfacility transports are at risk of disruption by AV companies?
8. Why are drone deliveries gaining traction in both rural and urban EMS settings?

   a. They eliminate the need for skilled medics
   b. They offer a lower cost than air ambulances
   c. They can deliver time-sensitive supplies faster and more directly
   d. They bypass FAA regulations

9. Which U.S. government agency provides a flexible framework for deploying automated driving systems and calls on emergency services to help shape policy?
10. What leadership actions can EMS agencies take now to avoid being displaced by external autonomous transport providers?

CHAPTER SUMMARY

Drones, Driverless Tech, and the Future of Emergency Medical Services Mobility

AVs and drones are no longer science fiction—they are active, expanding, and already serving millions of people in urban, rural, and healthcare environments. From Waymo's driverless taxis to Rwanda's drone-based blood delivery, autonomous systems are proving they can move people, supplies, and information more safely and efficiently in ways EMS has yet to embrace fully.

This chapter examined how these technologies intersect with EMS operations, from the rise of autonomous ambulances like Volkswagen's ID. Buzz, to the deployment of Beep Mobile's medical shuttle programs in Peoria, Arizona. We explored the workforce implications, the urgent need to modernize state regulations, and the critical safety comparisons between human-driven ambulances and AVs.

We also looked forward to the growing $50 billion ambulance transport market, the financial realities of nonemergency medical transport, and the potential for external disruption if EMS fails to lead. The lesson from taxis is clear—those who resist innovation lose control of their future.

EMS cannot afford to wait.

Autonomous transport systems offer EMS a chance to solve long-standing challenges—staffing shortages, safety risks, rural coverage gaps, and financial fragility. However, that opportunity will only be realized if EMS leaders act now. Build the partnerships. Reform the regulations. Start the pilots. Shape the future—before someone else does.

10 Artificial Intelligence–Enhanced Emergency Medical Services Clinicians

LEARNING OBJECTIVES

After completing this chapter, the reader will be able to

- Explain the concept of cognitive load and its implications for EMS clinicians, including its role in diagnostic error, clinician burnout, and task-saturated environments.
- Differentiate between intrinsic, extraneous, and germane cognitive load, and assess how these categories manifest during EMS operations.
- Describe how AI-powered tools can support EMS decision-making, including real-time risk detection, protocol adherence prompts, and clinical insight augmentation.
- Apply the aviation concept of autopilot to EMS practice, illustrating how task-sharing and system support can enhance safety and performance without diminishing clinician authority.
- Identify the impact of cognitive overload on documentation accuracy, medication safety, and situational awareness, and describe how AI can help reduce these risks.
- Discuss the mental health and professional well-being benefits of real-time clinical support tools, including reduced second-guessing and enhanced clinician confidence.
- Evaluate the ethical boundaries of AI integration in EMS, including maintaining human presence, empathy, and clinician judgment at the center of care.
- Propose leadership strategies for adopting AI-enabled support systems, prioritizing clinician safety, cognitive relief, and high-performance decision-making.

I have lost count of how many trauma patients I have treated in the back of an ambulance. And while I am now approaching honorary membership in the "Dinosaurs of EMS" club, I still remember the feeling.

Whether it was a rural call with 45-minute transport times or the shoulder of a highway at 2 a.m., I have stood over multisystem trauma patients, often as the only paramedic on scene. Sometimes, I had a solid emergency medical technician beside me. Occasionally, I had the luxury of a second medic. But most of the time, it was just me—making decisions, pushing meds, managing an airway, starting lines, trying to piece together a clinical picture in the middle of chaos.

Like many of you, I have worked in rain, snow, and mud. Under highway lights and in ditches with barely any light at all. I have documented vitals on bloody gloves and made life-or-death decisions with limited data and even less support.

And I vividly remember what it felt like to walk into the trauma bay at the hospital.

Fifteen people would be waiting: a trauma surgeon, an emergency department physician, respiratory therapists, nurses, techs—maybe ortho and surgery already on standby. Suddenly, the solo medic caring for a critically injured patient was replaced by a coordinated ballet of trained professionals—each with a role, a monitor, a task.

That contrast hit me every time: Why was I expected to do what the rest of the medical system does as a team *alone*?

Do not get me wrong—I am proud of what we do in EMS. We're trained to improvise, adapt, and carry the load. But at some point, that expectation becomes a liability. The field is moving too fast. There's too much data, too many tools, and an overwhelming amount of clinical complexity to expect anyone to manage it all in isolation.

That is where this chapter begins.

Because for the first time in history, we have a way to change that. Not with more staffing—those days are gone—but with something just as powerful: technology.

This includes AI-powered decision support, augmented reality, heads-up displays that reduce cognitive load and flag subtle clinical changes before they are detected by clinicians, and more.

We cannot bring a trauma team into every ambulance, but we can get their insight, pattern recognition, clinical reasoning, and a second set of eyes.

Collaboration is not a weakness. It's what medicine does best. Collaboration is what physicians do in every trauma bay, cardiac catheterization lab, and intensive care unit nationwide. EMS clinicians need the same support without having to beg for it.

This chapter is personal because I know what it feels like to be out there, making decisions at 60 miles per hour with limited help. But I also know what's possible. The future of EMS includes giving *you*, the EMS clinician, something we have never had before: *superpowers* in the form of

- AI support that sharpens your judgment,
- advanced telehealth collaboration that brings specialists into the rig, and
- virtual assistants that monitor trends, catch what you cannot, and reduce cognitive overload.

We do not need to prove we can do it alone. We need to prove we are ready to do it better.

Let's talk about what that looks like.

Under Pressure: The Cognitive Load of Emergency Medical Services

EMS clinicians operate under one of the highest cognitive loads in medicine. We make critical decisions with limited information while multitasking, communicating, performing procedures, and documenting in unpredictable environments. Most of the time, we do it alone or with one partner. That level of mental juggling is not just impressive—it is risky.

Cognitive load theory breaks down the mental effort required into three types:

- **Intrinsic load:** the complexity of the task itself
- **Extraneous load:** unnecessary mental strain caused by poor systems, like fragmented electronic health records (EHRs) or unclear protocols
- **Germane load:** the beneficial processing that supports learning and decision-making

All three are high in EMS—and the stakes are even higher.

A national survey of over 5,200 U.S. physicians found that high cognitive task load, measured via the NASA-TLX[1] tool, was among the most potent predictors of clinician burnout,[2] more than work hours or patient load. Physicians in high-pressure specialties like family medicine and oncology reported burnout rates above 30%. While EMS was not included in that survey, emergency medicine clinicians share many of the same cognitive stressors and arguably face even more fragmentation and chaos.

According to the Agency for Healthcare Research and Quality, extraneous cognitive load—caused by interruptions, multitasking, poor EHR interfaces, and alarm fatigue—accounts for up to 87% of medical diagnostic errors.[3] That is not a typo. Most medical errors stem not from a lack of knowledge but from cognitive overload. If that happens in hospitals with teams and systems, imagine what it means in the back of an ambulance with one or two people, no lab access, and a noisy, moving vehicle.

Now add modern demands: complex medication regimens, sleep deprivation, backlogged documentation, dynamic protocols, hospital bypass decisions, and a growing list of patient expectations. This is the reality of EMS in 2025—and it's time to recognize it. Artificial intelligence does not erase this pressure; it redistributes it in ways that matter at the point of care.

For the future EMS clinician, when real-time AI systems help calculate and track the Shock Index, suggest differential diagnoses, or calculate drug dosages, it is not replacing you. It's lifting part of the load. It's doing what hospital systems have done for years—using tech to filter noise, highlight risk, and give clinicians space to think.

Cognitive bandwidth is a finite resource. Mistakes happen when you are maxed out—not because you are bad at your job, but because your brain is full.[4]

In EMS, we are taught to carry the weight ourselves—to make the call, push the med, and move on. But as the volume, complexity, and consequences of those decisions rise, the most innovative clinicians will not be the ones doing it all—they will be the ones who use the right tools to do it better.

Leveraging cognitive support tools is not cheating. It is medicine done right.

Workload, Well-Being, and the Case for Cognitive Relief

Cognitive overload often indicates a system's failure. When EMS clinicians are burned out, distracted, or making decisions in task-saturated conditions, patient care suffers. However, slogans about resilience or another wellness webinar do not address this. Cognitive overload is mitigated by designing systems that support cognitive effort rather than depleting it.

The American Medical Association (AMA) has highlighted cognitive ergonomics as one of the most underutilized tools in combating burnout and error across health care (fig. 10–1). Their guidance is blunt: In order to improve patient safety, reduce mistakes, and protect your workforce, you must stop designing systems that overload the clinician's brain. That means simplifying interfaces, consolidating data streams, reducing redundant alerts, and deliberately standardizing procedures in a way that supports—not stifles—clinical decision-making.

Sound familiar? It should. EMS has redundant documentation, split-screen distractions, fragmented protocols, and unnecessary cognitive hoops. The same tools being applied to protect hospital physicians must now be extended to EMS clinicians in the field. Cognitive relief is not a luxury—it's a safety issue. This must become a leadership priority if our EMS systems are serious about wellness, safety, and retention.

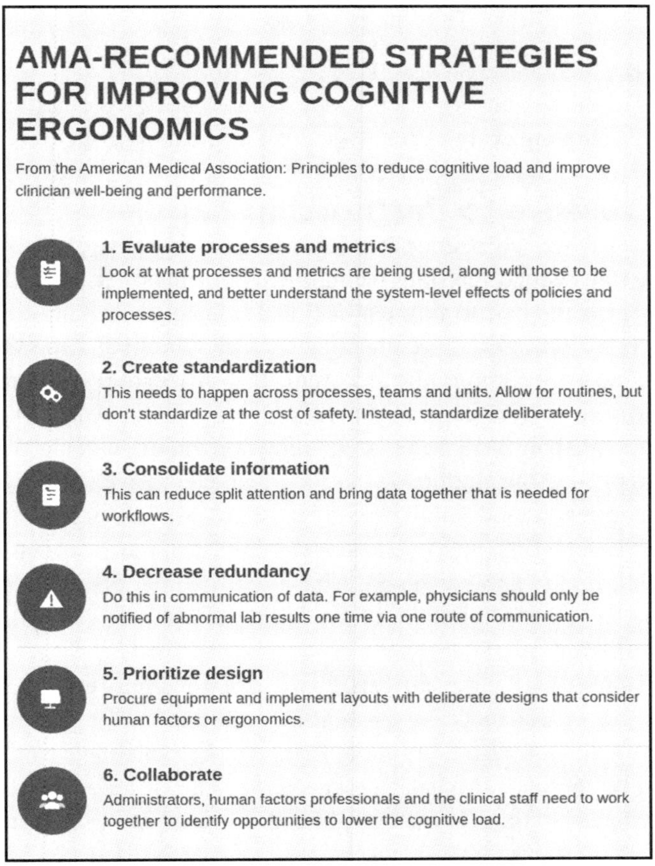

FIGURE 10–1. The AMA identifies six strategies to reduce clinician cognitive load: evaluating processes, creating standardization, consolidating information, decreasing redundancy, prioritizing design, and fostering collaboration. Together, these principles promote efficiency, safety, and clinician well-being.

Less Overload, More Bandwidth

Many of us in EMS have replayed calls, wondering what we missed:

- The patient who coded unexpectedly
- The rushed (and maybe even incorrect) medication calculation
- The protocol step you are not sure you completed because the scene was chaotic, the family was screaming, and you were doing the best you could with what you had

You have been there. So have I.

This is where the conversation about AI must shift. It's not about making us faster. It's about making us feel less alone in the most critical moments.

From Pressure to Partnership

EMS clinicians routinely operate at the outer edge of cognitive performance. You are making split-second decisions with limited information in dynamic environments—often while performing physical tasks, managing emotional families, coordinating with hospitals, documenting care, and thinking three steps ahead.

That is task saturation, and it is real. It's why errors happen. It's why second-guessing plagues good clinicians after tough calls. And it's why AI has a moral imperative in EMS: to act as a digital copilot, not to replace your judgment but to offload the clutter so your judgment can thrive.

I'm a paramedic and an airplane pilot. The parallels between EMS and aviation are fascinating, and I believe EMS can learn much from the aviation industry (but that is another book). I have learned that confidence does not come from being perfect. It comes from knowing you have systems and backups to support you.

When I am hand-flying an aircraft, I am focused and fully engaged. However, when the workload increases—weather, airspace changes, instrument approaches—activating the autopilot brings a sigh of relief. Not because I cannot fly, but because task-sharing keeps me safer. It reduces my cognitive load, allowing me to think bigger, stay ahead, and remain calm under pressure.

EMS is no different. There are moments when you must hand-fly every second of patient care. But there are also times when you need a system that catches what you missed, prompts you with a dose check, reminds you of the subtle electrocardiogram change, or flags a trend you had not processed yet.

That is not a weakness. That is leadership. Aviation and medicine work best in teams, with checks and balances.

Better Care Means Better Clinicians

Here's the secret: This is not just about improving patient care. It's about *you.*

AI-powered tools can help reduce burnout, second-guessing, and the nagging fear that you missed something important. Decision support validates your actions, flags early decompensation, or simplifies documentation, freeing your mind—and that has profound mental health benefits.

We have already seen this in physician practice. Chapter 5 discussed how AI tools embedded in EHRs cut documentation time and reduce doctors' cognitive fatigue. EMS deserves the same. Our clinicians deserve systems that support them on the scene, not just in the software.

The Human Element Still Matters Most

AI can guide dosing, catch deterioration, and support decisions—but it cannot offer a steady hand or a reassuring word in the back of an ambulance. That is your role, and always will be.

Empathy must remain nonnegotiable as EMS clinicians integrate these tools into their practice. A patient in crisis does not just need data-driven decisions—they need to feel safe, seen, and supported. AI cannot recognize fear in a child's voice or tension in a partner's glance during a domestic call, but you can.

Human judgment, compassion, and presence are irreplaceable. AI is here to augment—not erase—the heart of our work.

The Big Picture: Confidence, Not Control

AI in EMS is not about controlling the call. It's about reinforcing your confidence that you did not miss something critical.

It's about knowing that even in the worst circumstances—bad lighting, signal, vibes—you had another set of digital eyes on the problem. It's knowing that after the call, you did everything possible, and you did it right.

That is not automation. That is assurance.

EMS clinicians deserve that support because we have carried the burden alone for a long time.

Key Terms and Definitions

- **Agency for Healthcare Research and Quality:** A U.S. federal agency focused on improving the safety and quality of the American healthcare system. This agency identifies cognitive load as a key contributor to diagnostic errors.

- **AI-Powered Decision Support:** AI tools that analyze patient data in real time to provide clinical insights, suggest interventions, or flag potential deterioration without replacing the clinician's judgment.
- **Cognitive Ergonomics:** The science of designing systems, tools, and workflows to support mental performance and reduce cognitive burden, improving safety, efficiency, and clinician well-being.
- **Cognitive Load:** The total mental effort required to perform a task. In EMS, cognitive load includes decision-making, multitasking, emotional management, and environmental challenges.
- **Cognitive Load Theory:** A framework describing how the brain processes information under stress, dividing load into intrinsic (task complexity), extraneous (environmental distractions), and germane (learning-enhancing) categories.
- **Contemporaneous Evidence:** Documentation created at the time of care, carrying significant legal and clinical weight due to its timeliness and authenticity.
- **Decision Fatigue:** The decline in quality of decisions after extended periods of mental effort, every day in high-stakes, task-saturated EMS environments.
- **Digital Copilot:** A metaphor for AI systems that assist clinicians during patient care by managing background tasks, offering alerts, and surfacing critical information without taking control.
- **NASA-TLX (NASA-Task Load Index):** A validated tool used to measure perceived workload across several dimensions (mental, physical, temporal demand, performance, effort, frustration), often used in healthcare burnout studies.
- **Shock Index:** A clinical indicator calculated by dividing heart rate by systolic blood pressure to identify early hemodynamic compromise. AI tools can track and trend this metric in real time.

Reflection Questions

1. Think back to your last high-stakes call. Where would real-time clinical support have helped the most?
2. How would your confidence change if you knew AI was monitoring for clinical deterioration in the background?

3. What parts of your current EMS documentation, medication calculation, or triage process could be supported or improved with AI tools?
4. How does the aviation analogy of autopilot resonate with your experience as a clinician? When would task-sharing improve performance?
5. How should EMS leaders prioritize adopting AI tools that reduce burnout, improve accuracy, and support clinicians during critical decision-making?

Study Questions

1. Why does the chapter argue that traditional expectations of EMS clinicians—performing complex tasks alone under pressure—are no longer sustainable?
2. What is cognitive load, and how does it specifically impact EMS clinicians in the field?
3. According to the chapter, what percentage of medical diagnostic errors are attributed to extraneous cognitive load?
4. True / False: AI-powered tools are designed to replace EMS clinicians and automate critical care decisions.
5. What analogy from aviation does the author use to describe how AI can support EMS clinicians in high-stress environments?
6. Which of the following is *not* identified as a source of cognitive overload in EMS practice?

    a. Weather conditions
    b. Poor EHR interfaces
    c. Redundant protocols
    d. Alarm fatigue

7. The AMA recommends using ______ ergonomics to reduce burnout and medical errors.
8. Why is the concept of AI as a copilot emphasized in this chapter?
9. What are two ways AI can help reduce the mental burden on EMS clinicians during a call?
10. What does the chapter suggest is the ultimate goal of AI integration in EMS: control, speed, or confidence?

CHAPTER SUMMARY

Lifting the Load, Elevating the Clinician

For decades, EMS clinicians have been asked to do what no other medical professional is asked to do—operate alone, in chaos, without support, without context, and with room for error.

We have carried that burden proudly, but the world has changed.

Cognitive load in EMS is now extreme, driven by growing complexity, multitasking demands, chaotic scenes, limited staffing, and fragmented technology. This is not just exhausting; it's unsafe.

AI does not replace the EMS clinician—it protects them. It distributes the load. It flags risks early. It validates your judgment and reduces the chance of errors when you are task-saturated or cognitively maxed out.

AI-powered tools can calculate a Shock Index in real time, suggest medication adjustments, recognize subtle electrocardiogram changes, or prompt protocol adherence—all without interrupting patient care. This is not about automation. It's about assurance.

Like aviation, where autopilot does not replace the pilot, it empowers them. AI in EMS is about giving you the bandwidth to think clearly, act decisively, and recover confidently after each call.

The future of EMS will be built on clinical excellence, cognitive support, and compassionate care. Clinicians like you will make it—ready to do it better, not alone.

11

Ethics, Accountability, and the Emergency Medical Services Artificial Intelligence Framework for Action

LEARNING OBJECTIVES

After completing this chapter, the reader will be able to

- Explain why ethical AI integration in EMS begins with data integrity, and describe the risks of using incomplete, inconsistent, or biased datasets in training AI models.
- Identify how bias manifests in healthcare AI tools, using real-world examples from hospital and EMS settings, and assess how EMS documentation practices may unintentionally contribute to bias.
- Describe the importance of explainability in AI-supported clinical decisions, and outline how EMS clinicians should respond to AI-generated recommendations that conflict with their observations or clinical judgment.
- Evaluate the privacy risks associated with AI use in EMS, particularly regarding data centralization, pattern recognition, and compliance with HIPAA and other regulations.
- Discuss the legal and professional responsibilities of EMS clinicians when using AI, including the need for documented clinical reasoning and the risk of deferring uncritically to automated outputs.
- Apply national standards, such as the National Institute of Standards and Technology (NIST) AI Risk Management Framework, to guide the ethical implementation of AI tools in EMS agencies and software platforms.
- Define the shared responsibilities between EMS clinicians, vendors, medical directors, and regulators in ensuring transparency, safety, and ethical alignment in AI deployment.
- Propose strategies for establishing accountability mechanisms, including audit trails, training protocols, and vendor disclosures, to mitigate risk and enhance trust in AI-supported EMS practice.

Prologue: Ethics in the Back of the Ambulance

You are 20 minutes out from the hospital with a 71-year-old Black female. She does not speak English, lives alone, and was found confused with borderline vitals. You do not have much to go on—just her name, a medication list with Eliquis and glipizide, and a family member who says, "She hasn't been right all day."

Your newly implemented AI-powered clinical decision support system flags her as a moderate risk. There are no sepsis or stroke alerts. The AI clinician decision support protocol suggests routine transport and standard notification.

But something feels off.

You are observing her vital signs trend downward. She is pale. Her mental status is fluctuating. You cannot explain it with just age or medications. She is unstable, and she is ill. And she needs help in a packed emergency department, not 45 minutes from now.

This is where ethics and AI collide—right there in the back of your ambulance.

In 2019, a landmark study published in *Science* uncovered that a widely used healthcare algorithm—one used to shape care plans for millions of patients—was deeply biased. At the same risk score, Black patients were significantly sicker than White patients. The reason? The algorithm had been trained to use healthcare costs as a proxy for illness. Because less money is historically spent on Black patients, due to systemic inequities, the algorithm incorrectly concluded they were healthier.[1]

In other words, the AI did not just inherit bias—it amplified it.

That is a warning for all of medicine, and certainly EMS.

If we integrate AI into prehospital care without serious scrutiny—without trained ethicists, transparent validation, and ongoing bias audits—we will automate injustice into our workflows. And it will be your patients who suffer.

AI is not the enemy, but it is also not neutral. It reflects the data it's fed, the assumptions it's built on, and the blind spots we fail to address.

As EMS systems incorporate AI, we must demand more: transparency, accountability, medical oversight. Because the consequences are not theoretical, they're on your stretcher.

The patients depend on you to get it right, whether the algorithm does or not. (*Note*: This chapter continues the ethical roadmap introduced in chapter 6. There, we outlined why EMS must lead in the AI era. Here, we deliver the framework and the ethical imperatives that must guide our actions.)

Ethics in the Real World: Three Moments That Should Give Us Pause

Ethical concerns in AI are not hypothetical. They surface in everyday EMS practice—in the ambiguity of triage, the urgency of transport decisions, and the documentation that may later serve as the only remaining call record.

Consider these three scenarios:

The Missed Voice

An AI-powered triage assistant in a 9-1-1 dispatch center downgrades a call from a non-English-speaking caller. Trained primarily on English-language calls, the algorithm misses key urgency cues. No one corrects it. The nearest advanced life support unit is rerouted. The delay proves fatal.

The Unequal Map

A predictive deployment system sends advanced life support units to high-frequency, high-income neighborhoods with a robust documentation history. Poorly represented in training data, lower-income areas are gradually deprioritized—not through intent but inertia.

The Incomplete Narrative

An AI-generated narrative documents: "Patient refused transport, alert and oriented." It omits the patient's verbal hesitation, the family's concern, and the cultural nuances of the interaction. Weeks later, in a complaint investigation, those details are unrecoverable.

These scenarios are not far-fetched. They are possible. In some systems, they may already be happening.

When AI tools are deployed without oversight, transparency, or ethical safeguards, the result is not just flawed care—it is codified injustice.

These moments are not presented to provoke fear. They are here to sharpen our vision. Each is a signpost—pointing to where ethical design failed, where inclusion was overlooked, and where leadership must step in. EMS does not need to be perfect to lead, but we must be present; present when systems are built, and present in conversations about fairness, transparency, and trust.

These cases do not close the door on AI. They challenge us to open it more wisely—with clarity, courage, and community insight.

Ethics in EMS AI do not begin with the algorithm. They start with the clinician, who must see what the machine cannot.

Artificial Intelligence Ethics Begins with Data Integrity

You cannot build ethical AI on unethical data.

The EMS profession is fortunate to have one of the most expansive, standardized clinical datasets in health care: NEMSIS, the National EMS Information System. It powers national surveillance, informs federal policy, and feeds nearly every AI-related EMS research model being developed today.

However, having data is not the same as having good data, and this distinction is foundational to ethically integrating AI in EMS.

Remember all those drop-down boxes you have selected while writing patient care reports? The blood pressure category, the impression list, and the medication administered field? Those are not just compliance clicks anymore. They are the building blocks for every EMS-specific AI model being developed for clinical decision support. Now, more than ever, they must be accurate, complete, and meaningful.

As Samta Kapoor, Ernst & Young's AI Leader points out, "AI absorbs everything it receives without judgment, so individual user choices can easily affect AI outcomes, becoming a permanent part of the AI universe."[2]

This includes every checkbox, shortcut, and incomplete narrative. Once it enters the dataset, it's part of the training signal that guides future decisions—not just yours but across the system.

In 2021, peer-reviewed analysis of Michigan's EMS data showed staggering variability and gaps in critical clinical fields. Among over 12 million patient care records, researchers found that some of the essential fields for AI model training—like vital signs, medications, interventions, and provider impressions—were missing more than 90% of the time in specific agency submissions.[3]

From an AI development perspective, this is potentially a significant issue. If a dataset is incomplete, unevenly coded, or lacks clinical depth, any AI system trained on it is already compromised. It will reflect—not reality—but the limitations, biases, and documentation inconsistencies embedded in the data.

This is not limited to Michigan. These challenges exist in varying degrees across every state.

Ethics in EMS AI system design starts with data governance:

- Clear data standards and mandatory field validation
- Transparent data provenance from field to database
- Clinician-friendly documentation tools that capture what matters, without adding noise

You cannot fix biased algorithms by tweaking outputs. You fix them by building ethical inputs—clean, complete, context-rich data that genuinely represents the patients we serve.

If we do not get this right, every prediction that follows—every sepsis alert, risk score, or triage decision—will echo our inattention.

Ethical AI starts with ethical data, and that responsibility is ours.

Bias Is Not Just a Possibility—It Is a Proven Risk

In the prologue, we have already seen how a flawed AI design—trained on cost rather than clinical need—nearly eliminated Black patients from critical care coordination systems. That was just one system; however, the truth is that it is not alone.

In 2025, researchers in France trained a generative AI model on over 480,000 emergency department triage notes.[4] When they changed only the patient's gender, leaving the clinical picture unchanged, the AI's triage decisions changed, too. Women were more likely to be under-prioritized, and men were slightly more likely to be over-triaged. The bias worsened when the AI mimicked junior clinicians' documentation patterns.

That should hit close to home for EMS. Our data is often incomplete, entered under duress, and sparse in context. In other words, it is a potential bias amplifier waiting to happen.

Bias in AI does not mean someone wrote racism into the code. It implies the algorithm learned from a system that treats patients unequally and assumed that was correct. That is the danger. AI will replicate what it sees. If our data carries bias, our algorithms will have bias. If our documentation lacks clinical depth, our AI will lack clinical accuracy. If we ignore health equity, AI will codify health inequity.

The French triage study clarified something else: Bias hides in patterns, not people. When the system was retrained on a more diverse dataset and tested against a balanced clinical validation set, performance and fairness improved. It was not a magic fix, but it was measurable progress.

It's our job now. EMS cannot unquestioningly adopt hospital-based AI models or national datasets without scrutinizing them for equity, accuracy, and clinical context. Our protocols must be reviewed through a health equity lens, and our models must be validated on our patients, not just urban academic datasets.

Bias in medical AI systems is unfair and dangerous. If EMS systems do not actively challenge it, we will make and systematize mistakes.

Explainability, Trust, and the Human Clinician

There is a reason patients trust people more than machines, especially in moments of crisis. The look in your eyes, the calm in your voice, and the way you kneel beside them instead of standing over them are not algorithms. They are human. And in EMS, they are often the only thing holding the scene together.

That is why explainability matters.

As we integrate AI into EMS clinical decision support systems, we cannot afford to treat these tools as black boxes. If the system flags a sepsis risk or triage priority, you must understand why. Suppose you cannot explain the reasoning behind an AI-generated alert to your patient, partner, medical director, or emergency department physician. In that case, the tool has no business guiding care.

Explainability is not a tech problem—it is a trust problem.

According to the *American Medical Association Journal of Ethics*, patients have a right to understand how decisions are made about their care, even when AI assists those decisions.[5] If a clinical support system suggests a diagnosis or intervention and you cannot articulate its rationale, you ask patients to trust something invisible. In EMS, where trust must be earned in seconds, it is a liability.

Clinicians need clarity, too.

A recent study by Cansu Elgin and Ceyhun Elgin emphasized that AI used in acute settings must be designed with human interpretation in mind.[6] EMS clinicians must know where the data came from, the logic that is used, and when to question the result. Anything less than that erodes your confidence and can degrade patient safety.

EMS AI tools must be interpretable by design—accurate and transparent. Explainability is not optional. It must be embedded into every AI system used in EMS. Alerts must include contextual reasoning. Contributing factors must accompany risk scores, and clinicians must be trained to accept and challenge the output when necessary.

A cornerstone of implementing AI ethically and responsibly is remembering that AI is not accountable. You are. EMS clinicians must remain the final decision-makers on the scene. The job of AI is to inform, not to replace.

Protecting Privacy, Preserving Trust

AI is only as trustworthy as the data that powers it. In EMS, we collect some of the most sensitive data in health care.

EMS clinicians document medical, behavioral, and social details every shift—often in the middle of chaotic, high-stakes situations. These records do not just reflect medical events; they reveal living conditions, mental health crises, substance use, and personal histories. This is deeply personal information, captured without a front desk, intake forms, or consent signatures.

As we integrate AI into EMS systems, data privacy and patient trust cannot be afterthoughts. They must be built in from the start.

The Risk Is Not Theoretical

Across the healthcare industry, breaches have exposed sensitive patient data, eroded trust, and created real harm. In EMS, the risks are amplified by fragmented systems, third-party vendors, and inconsistent data security practices.

AI changes the game. Once trained, these systems do not just process data—they retain patterns. Even when identifiers are removed, advanced models can sometimes infer personal details, raising serious concerns about compliance with the HIPAA, the General Data Protection Regulation, and emerging state privacy laws.

Privacy by Design

Responsible EMS leaders must prioritize privacy by design—integrating safeguards into every layer of AI development and data exchange. That means

- encrypting all EMS data (at rest and in motion),
- exploring federated learning to reduce data centralization risks,
- enforcing granular access controls with full audit logs, and
- implementing transparent patient data policies that explain how EMS information is used, stored, and protected.

Trust begins with transparency. Patients who believe their data is safe are more likely to engage with care. Those who fear misuse are more likely to withhold critical information or avoid care altogether.

If we want AI to strengthen EMS, it must start with protecting the very people we serve. Because in this next chapter of EMS, trust will be just as important as technology.

The Clinician Is Still in Charge

AI can process faster. It can remember everything. It can find patterns no human would ever notice. But AI cannot take responsibility for the patient in front of you. That is still your job, and it always will be.

EMS clinicians do not get to abdicate accountability just because a machine suggested a diagnosis. As EMS clinicians, we remain the final filter, whether it's a sepsis risk score, a stroke pre-alert recommendation, or a medication dose calculation. That is not a limitation—it's the safety net. It's what protects patients from false positives, missed context, and the real-world variability no algorithm can fully grasp.

Let's be clear about what AI is—and what it is not. AI is not a clinical license. It's not a replacement for judgment. It's not a scapegoat when something goes wrong. As EMS agencies begin integrating clinical decision support tools, we must ensure that every clinician understands that they are not expected to follow AI. You are expected to lead.

This is where training matters.

A well-designed EMS AI tool must not just produce outputs—it must explain its reasoning. If the algorithm flags sepsis, you should be able to see why. Was it heart rate variability? Temperature drift? A borderline modified shock index? As EMS professionals, we should know how to document our reasoning, both when we follow AI recommendations and when we choose a different course.

This is called explainability, and it's a cornerstone of ethical AI.

But even with that, you must be trained to ask, "Does this make sense in the real world?" If the AI flags a patient as low-risk, but their skin tone is pale, their breathing is shallow, and your gut screams, "This is not right," then trust your training. As EMS leaders—clinicians, educators, and decision-makers—we must trust our training and professional judgment. AI is not accountable for our actions—we are.

Here's the truth: There's growing legal and professional risk in ignoring clinical decision support. As AI gets smarter, there will be cases where not using it, or ignoring its insights, could be questioned. But the reverse is just as true. If a poor outcome occurs and the clinician simply defers to a flawed or biased algorithm, that will not hold up in court. Nor should it.

This is not about following AI—it's about collaborating with it.

This model is already being deployed at Cedars-Sinai, a hospital system in Los Angeles, California, that cares for more than 1 million people annually across over 40 locations. AI is leveraged to augment rather than replace human expertise. [7] They call it "human in the loop." The AI offers input, but the clinician retains authority. That is how it must work in EMS, too—especially in high-stakes, low-resource environments.

We must normalize three things:

- Reviewing AI recommendations
- Challenging them when necessary
- Documenting your clinical rationale, especially when you override AI

This is not new. Physicians already do it with telemetry, radiology reads, or even medical control. EMS must adopt the same standards.

AI in EMS is powerful, but only when paired with professional judgment, clear policies, and clinicians trained to think, not just click. We do not want blind obedience to the algorithm. We want informed use, skeptical evaluation, and professional accountability.

No matter how advanced AI becomes, it will never replace your judgment, instincts, or responsibility.

Vendors and Systems Must Be Held Accountable, Too

EMS clinicians may carry the patient, but they should not have the entire liability for systems they do not control.

As AI becomes embedded in EMS documentation, dispatch, and clinical decision support, we must expand our understanding of accountability. This is not just a clinician's issue. It's a system-wide obligation that includes vendors, developers, regulators, medical directors, and EMS agencies.

Vendors must disclose model design, limitations, and training data assumptions if an AI tool was trained primarily on urban academic hospital data or was never validated in a prehospital environment; this must be made clear. EMS clinicians shouldn't be left guessing whether their tool is appropriate for their setting.

Medical directors must evaluate and approve AI systems like any other protocol change. These tools should not be treated as passive software—they influence patient care and must be reviewed accordingly.

State EMS offices and national regulatory bodies must set minimum standards for validation, transparency, explainability, and clinical oversight. AI tools cannot enter the field without a pathway for redress, correction, or deactivation if harm is suspected.

EMS leaders must hold the line. Do not let marketing hype override clinical reality. Do not deploy an AI tool if it cannot show its work, justify its conclusions, or withstand scrutiny. Our profession is not a beta test site for unproven technology.

The AI revolution in health care is real, but the risk of handing over clinical influence to unregulated, unaudited, and unaccountable systems is also real. Clinicians are responsible for decisions, but systems must be accountable for the tools they hand us.

If we fail to set that standard now, the consequences will not be theoretical. They will be measured in missed alerts, biased recommendations, and patients who should have received better care.

Ethical AI in EMS is not just about the user.

It's about the system behind the screen.

Anchoring Emergency Medical Services Artificial Intelligence Ethics in National Standards: The Role of the National Institute of Standards and Technology

Ethical integration of AI into EMS is not just a matter of best practice—it is a matter of national alignment.

In 2023, the National Institute of Standards and Technology (NIST) released the AI Risk Management Framework[8] to guide the development of safe, trustworthy, and transparent AI systems across all sectors (fig. 11–1). That same year, NIST launched the U.S. Artificial Intelligence Safety Institute—a coalition of over 200 organizations dedicated to ensuring that next-generation AI systems are rooted in science, ethics, and public trust.

For EMS, these efforts are not peripheral—they are foundational. EMS agencies and vendors must explicitly align with NIST's risk-based approach to ensure AI systems are effective, resilient, and ethically grounded. The AI Risk Management Framework provides a clear, adaptable structure for designing and governing AI systems that interact with human lives in real time, which is exactly the environment in which EMS operates daily.

NIST's AI Risk Management Framework is built around four core functions:

- **Govern:** Establish organizational policies, structures, and accountability systems that define how AI-related risks are identified, tracked, and mitigated.

- **Map:** Understand the context in which AI will operate, including system design assumptions, intended users, societal impacts, and EMS-specific environments.
- **Measure:** Evaluate AI systems for validity, fairness, robustness, and bias—using quantitative and qualitative tools to assess performance in EMS-relevant conditions.
- **Manage:** Proactively address risks throughout the AI lifecycle; update systems, adjust policy, and continuously learn from user experience, clinical data, and community input.

These are actionable standards that EMS agencies can adopt today.

For example, *governance* means not just writing policy, but ensuring EMS agencies have structures to pause or revise AI systems if they degrade over time. *Map* means evaluating how a sepsis prediction tool trained on urban hospital data might perform in a rural EMS agency. *Measure* means developing clinician-facing metrics for accuracy, bias detection, and user trust. *Manage* means adapting workflows as frontline clinicians provide feedback on alerts, documentation burden, or false positives.

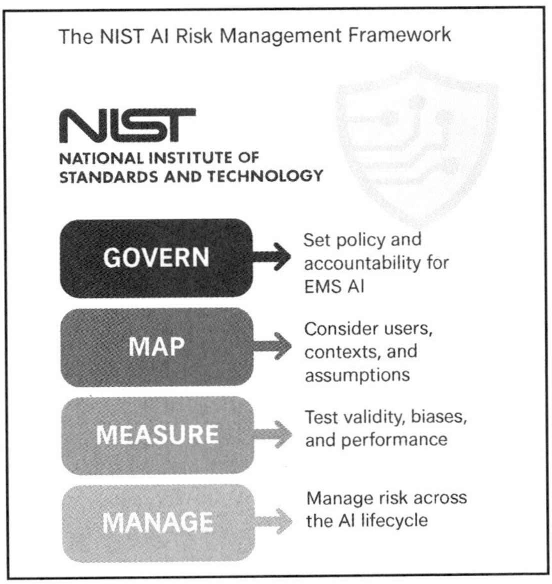

FIGURE 11–1. The NIST outlines four pillars for safe and reliable AI adoption: govern, map, measure, and manage. This framework guides EMS leaders in establishing policy, assessing bias, validating performance, and managing risk throughout the AI lifecycle.

Most importantly, NIST makes clear that trustworthy AI systems must be

- valid and reliable in high-stakes environments,
- accountable and transparent in logic and application,
- explainable and interpretable by end users,
- privacy-preserving and secure, and
- fair and inclusive for all populations served.

These values mirror the core of EMS practice: patient protection, clinical accountability, and public trust. EMS can lead, not lag, by aligning with NIST in setting the national bar for safe, ethical AI integration.

Ethical Principles for Integrating Artificial Intelligence

The table below aligns EMS's practical realities with the NIST AI Risk Management Framework and this chapter's broader ethical lessons. These principles provide a clear roadmap for EMS agencies, vendors, educators, and regulators seeking to implement AI responsibly, equitably, and clinically effectively (table 11–1).

TABLE 11–1. Ethical Principles for Integrating AI

Principle	Description
Patient Privacy and Data Security	Ensure robust protection of patient data through compliance with HIPAA and other regulations. Implement advanced encryption and security protocols.
Transparency and Explainability	Use AI systems that provide clear, understandable explanations for decisions and recommendations to ensure transparency.
Algorithmic Fairness and Bias Mitigation	Regularly audit AI algorithms to identify and eliminate biases; train on diverse datasets to prevent systemic bias.
Clinician Oversight and Final Decision-Making	EMS clinicians must remain the final decision-makers, using AI as a supportive tool; train clinicians to evaluate AI insights critically.
Continuous Education and Training	Develop ongoing education programs to keep clinicians updated on AI technologies; emphasize both technical and ethical training.

Principle	Description
Patient-Centered Care	Use AI to enhance, not replace, empathetic care; communicate clearly with patients about AI's role and ensure informed consent.
Ethical Use and Deployment	Establish guidelines for ethical AI use focused on patient well-being and equity; avoid AI use that could harm patients.
Collaboration with Stakeholders	Engage diverse stakeholders to inform AI integration strategies; foster interdisciplinary collaboration.
Accountability and Liability	Define clear accountability structures; promptly address AI-related errors with reporting mechanisms.
Innovation and Flexibility	Encourage ethical innovation; remain flexible and reassess AI tools to meet evolving healthcare needs.
Regulatory Compliance and Standardization	Comply with national and international AI regulations; standardize AI practices across EMS for consistent care.
Community Engagement and Trust Building	Engage with the community to build trust in AI use; ensure clear understanding of benefits and limitations.

Ethical AI in EMS is not a future aspiration but a present responsibility. These principles are safeguards and strategic enablers that allow innovation to serve patients, not bypass them. As EMS leaders, we must ensure that every deployment of AI reflects our values, respects the complexity of care in the field, and earns the trust of both clinicians and communities. When we embed transparency, fairness, and clinician oversight into the systems we build, we do more than manage risk—we lead purposefully. And in doing so, we set the standard for technology adoption and what ethical, human-centered EMS should look like in the next era (fig. 11–2).

Emergency Medical Services Artificial Intelligence Framework: A Draft for Action

Integrating AI into EMS presents unprecedented opportunities and significant challenges as the healthcare landscape evolves. To navigate this transformative era effectively, EMS leaders, educators, regulators, and clinicians must come together to establish a clear framework for action.

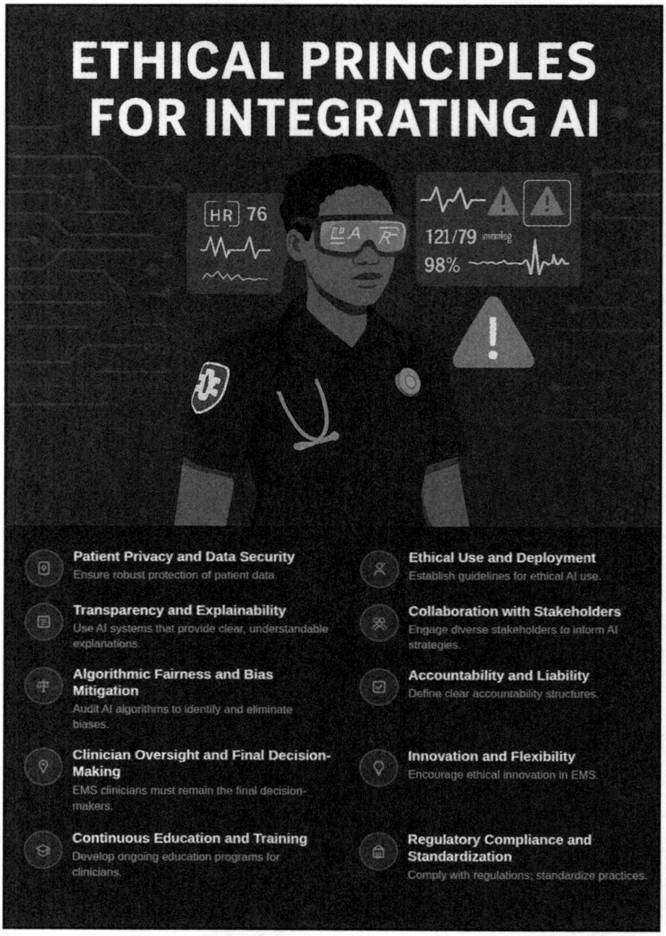

FIGURE 11–2. Integrating AI into clinical care requires more than technical expertise—it demands ethical discipline. Protecting privacy, ensuring transparency, mitigating bias, and maintaining clinician oversight are crucial for the deployment of trustworthy and responsible AI in EMS.

The following draft outlines essential steps for harnessing AI to enhance clinical practice while safeguarding the integrity of our profession. By adopting these principles, we can ensure that AI is a powerful ally rather than a replacement, fostering collaboration and innovation within the EMS community:

1. **Augment, not replace, EMS clinicians:** AI must be positioned as a support system—never a substitute—for clinical decision-making. Tools

must respect EMS clinicians' autonomy, training, and situational judgment.

2. **Unify the national voice:** National EMS organizations— the National Association of State EMS Officials, National Association of EMTs, National EMS Management Association, National Registry of EMTs, International Association of Fire Chiefs, American Ambulance Association, and others—must collaborate to publish joint guidance on AI use, policy alignment, and ethical standards. This includes formal position papers and readiness checklists.
3. **Educate the field:** We must immediately begin integrating AI literacy into EMS education. From documentation tools to diagnostic support and predictive models, EMS clinicians need foundational knowledge of how AI works, its risks, and how to assess its outputs critically.
4. **Advocate for representation at the federal level:** EMS must be formally included in federal AI policy development. From White House advisory panels to Health and Human Services initiatives, EMS should have a seat at the table—because the decisions being made now will impact our future viability as a profession.
5. **Guard against silent encroachment:** If EMS remains silent, others will shape the future of prehospital care without us. That future may be designed by voices unfamiliar with our clinical realities, operational needs, or regulatory structure. Technology companies, venture capital, logistics platforms, and AI developers are already building systems that affect emergency response. Without active EMS leadership, these systems may be deployed without clinical validation, local context, or alignment with EMS values.

Innovation without EMS input risks producing misaligned, ineffective, or even harmful tools. We must engage now to ensure AI is developed with—not just applied to—our profession.

These risks are already unfolding in adjacent industries. By proactively addressing these challenges, we can secure a future where AI catalyzes growth, innovation, and enhanced patient care within EMS.

Toward a Framework for Ethical and Responsible Artificial Intelligence in Emergency Medical Services

Despite meaningful progress across other branches of health care, EMS remains without a formal, profession-wide framework for guiding the ethical, effective, and responsible integration of AI. While robust frameworks require national collaboration, stakeholder consensus, and thoughtful governance, EMS cannot afford to delay its leadership role. We do not need perfection to begin—we need direction.

To assist with closing this gap, I'm proposing the *Conceptual Framework for Ethical and Responsible AI* in EMS. This is a clinician-first blueprint inspired by the work of the American Medical Association, American Nurses Association, American College of Physicians, and American Academy of Family Physicians. These national medical organizations have already built structured, principle-based models that clarify how AI can be safely and ethically deployed. Each offers a common conviction: AI must support clinicians, not sideline them.

Yet EMS is largely absent from this conversation. There is no national guidance on how AI should function in EMS documentation, dispatch, or decision support. No unified vision for how to protect the clinical role of emergency medical technicians and paramedics in an AI-enabled system. No clear standards for equity, transparency, or clinician education.

This absence is a growing liability. Without an EMS-led framework, others will define the future for us, and the results may not reflect our values, workflows, or mission.

It is time to change that.

A framework does not begin with regulation. It starts with shared values. The following model proposes twelve foundational pillars (table 11–2). It reflects what EMS professionals must demand of AI systems and what our patients should expect of us in return. It is not a regulation, but it is a rallying cry.

And it rests on one nonnegotiable truth: AI must enhance EMS clinicians, not replace them.

While many of the framework pillars outlined here align closely with those established by our physician and nursing colleagues, EMS operates in a fundamentally different environment. We deliver care outside clinical walls, often alone or in pairs, without access to a medical record, and with no opportunity to "come back later" to clarify documentation or consult a specialist. Decisions are made in motion. Records are written after the fact. And the settings—homes, highways, stairwells, and parking lots—rarely resemble the controlled environments assumed in most AI model development.

That is why two additional pillars—continuity and fidelity—have been added to this EMS-specific framework.

Continuity is clinical safety.

AI systems must enable seamless data sharing across the care continuum. EMS clinicians must be able to retrieve medical histories, medication lists, allergies, and care plans from health information exchanges (HIEs) in real time—before patient contact, not after. Likewise, EMS documentation must be transmitted to receiving facilities before arrival, not hours later or hidden

TABLE 11–2. Twelve Foundational Pillars

Framework Pillar	Core Commitment	Action in Practice
1. EMS Clinician Oversight	EMS clinicians retain full authority over patient care. AI must augment—not override—clinical judgment or field decisions.	Protocols must clarify when AI recommendations require review, documentation, or clinician override.
2. Field-Validated Performance	AI tools must be tested using EMS-specific data in real-world, prehospital settings—not only hospital-based simulations.	Require validation across urban, rural, and low-resource EMS environments before broad implementation.
3. Health Equity and Bias Accountability	AI must be actively evaluated for fairness, inclusiveness, and unintended harm across all populations and communities served by EMS.	Mandate demographic audits, require transparency on training datasets, and implement post-deployment equity monitoring.
4. Explainability and Clinician Clarity	EMS clinicians must understand AI outputs well enough to explain and act on them in the field. Black-box algorithms are not acceptable in high-risk care.	AI alerts must show rationale, contributing factors, and confidence scores, enabling transparent clinical communication and documentation.
5. Transparent Logging and Oversight	AI decisions, outputs, and clinician overrides must be logged and traceable for medical oversight, quality assurance (QA), and legal accountability.	Require system logs for every AI-driven recommendation, including whether it was accepted, modified, or rejected—and why.
6. Human-Centered System Design	AI tools must be designed for the realities of EMS—mobile, fast-paced, cognitively demanding, and often delivered by small teams in uncontrolled settings.	Include EMS clinicians in the design and user testing process; prioritize intuitive interfaces, hands-free options, and minimal cognitive load.
7. Interoperability and Open Innovation	EMS systems must support open architecture and interoperability with HIEs, computer-aided design, electronic patient care recording, and hospital electronic health records to allow responsible innovation and continuity.	Require open application programing interfaces, data exchange standards, and eliminate vendor lock-in that inhibits clinical and operational coordination.

(*continues*)

TABLE 11–2. Twelve Foundational Pillars (*Continued*)

Framework Pillar	Core Commitment	Action in Practice
8. Continuous Learning and Feedback Loops	AI tools must support post-deployment evaluation, include mechanisms for field feedback, and be subject to ongoing performance improvement.	Measure outcomes like false alert rates, missed flags, impact on clinical decision-making, and satisfaction—then iterate.
9. Community Engagement and Public Trust	EMS must build and maintain public trust by being transparent about how AI tools are used in emergency care.	Publish plain-language summaries, engage community groups, and include patients in governance or oversight discussions when appropriate.
10. Profession-Led Governance	AI policy, education, and standards in EMS must be shaped by national professional organizations representing the entire field.	Establish a national EMS AI council with stakeholders from NAEMT, NASEMSO, NEMSMA, NREMT, AAA, and EMS educators, to develop joint guidance.
11. Continuity of Care and Interoperability	AI systems must ensure continuity by connecting EMS to HIEs, hospital systems, and other providers before and after the call.	EMS clinicians must have access to medical history, meds, and allergies in real time—and hospitals must receive EMS records before the stretcher arrives.
12. Fidelity to Field Conditions	AI-generated documentation must reflect not just the clinical facts, but the environmental and situational realities unique to EMS care.	Narrative systems must capture the scene, response context, environmental risks, and barriers encountered—not just diagnosis or interventions.

Note. HIEs = health information exchanges, NAEMT = National Association of Emergency Medical Technicians, NASEMSO = National Association of State EMS Officials, NEMSMA = National EMS Management Association , NREMT = National Registry of Emergency Medical Technicians, AAA = American Ambulance Association

behind technical barriers. When EMS data is isolated, fragmented, or delayed, patients suffer, clinicians are blindsided, and system coordination breaks down.

Real-time interoperability is not a future vision—it is a safety requirement.

Emergency Medical Services–Specific Narratives

Beneficial AI does not stop at generating smoother sentences or autocompleting SOAP (structured subjective, objective, assessment, and plan format) notes.

In EMS, documentation is not just a summary—it is evidence, accountability, and a record of care delivered in the most uncontrolled environments in health care. AI must help emergency medical technicians and paramedics quickly and accurately preserve the full context of prehospital medicine.

The cluttered room; the chaotic scene; the safety and security threat; the noise, confusion, and emotional charge; the social dynamics; the barriers to care: These are not secondary observations. They are clinical indicators. They shape risk. They influence decision-making. They explain why actions were taken—or not.

Narrative fidelity is not a luxury—it is protection. It allows an EMS clinician's judgment to be understood days, weeks, or even years later. It turns a call into a clinical event that fits into a broader continuum of care. It protects clinicians from litigation, supports QA, and enables real-time hospital coordination.

This is one of the clearest examples of why AI-powered EMS documentation tools must fundamentally differ from those designed for clinics, hospitals, or other mobile health units. These systems must not simply replicate existing electronic health record workflows—they must be purpose-built for EMS, grounded in field realities, informed by the voice of EMS clinicians, and that nuance matters.

This is precisely why AI in EMS must be led by EMS, not simply handed down from external developers, hospital systems, or adjacent professions. If we do not shape the tools, we risk being shaped by them. And if we fail to demand systems that reflect our unique environment, we will inherit ones that erase it.

Ethical Foundations for Artificial Intelligence in Emergency Medical Services: Principles That Must Guide Every Tool

Ethics cannot be an afterthought. In a field defined by public trust, accountability, and real-time patient care, integrating AI into EMS demands ethical clarity from the very start.

These foundational principles are nonnegotiable guardrails for any AI tool that enters the EMS environment (fig. 11–3).

FIGURE 11–3. AI in health care must uphold the same moral foundations that guide clinical practice: autonomy, beneficence, nonmaleficence, justice, transparency, and accountability. These principles ensure that innovation never outpaces ethics.

Foundational Ethical Principles

- **Patient autonomy:** AI must never override a patient's right to informed decision-making. Where appropriate, patients should be made aware when AI is used, and EMS clinicians must always honor patient preferences.
- **Beneficence:** AI's goal is to improve health outcomes and enhance patient care. If it does not improve care, it does not belong in EMS.
- **Nonmaleficence:** AI must not cause harm, directly or indirectly. This includes avoiding decisions that perpetuate bias, compromise privacy, or mislead clinical judgment.

- **Justice:** AI must work for all communities. It must reduce, not reinforce, health disparities. Equity must be a requirement, not an aspiration.
- **Transparency:** EMS clinicians must understand how AI tools arrive at their recommendations. Patients and oversight bodies deserve clarity, not black boxes.
- **Accountability:** Every AI tool must have a transparent chain of responsibility. If something goes wrong, we must know who is accountable—and how it will be corrected.

Ethical Guidelines for Implementation

The principles above must translate into operational action. What follows are guidelines for EMS leaders, clinicians, developers, and educators at every phase of AI integration:

- **Design and Development**
  - **Data quality and governance:** Train models on accurate, diverse, and EMS-relevant data (poor data will produce poor tools—and put patients at risk); also, ensure privacy protections (e.g., HIPAA compliance) and establish governance processes to prevent data drift, loss, or misuse.
  - **Algorithmic fairness:** Require demographic representation in training data. Test for bias across race, geography, language, and socioeconomic status—and take corrective action when disparities appear.
  - **Explainability and clinical interpretability:** No AI system should be deployed if clinicians cannot understand its rationale. Field tools must show why a decision was made and allow EMS clinicians to act—or override—based on real-world judgment.
  - **Safety and reliability:** Test AI in live EMS environments, not just labs. Validate tools under stress, during peak call volume, and in diverse geographic conditions. AI that fails under pressure is not ready for EMS.
  - **Human oversight by design:** The clinician must always retain final authority. Systems must be designed with override capabilities and protocols that reinforce critical thinking, not passive automation.
- **Deployment and Use**
  - **Informed consent and communication:** Patients should know when AI is guiding elements of their care. EMS agencies should build a plain-language verbal disclosure into the encounter

workflow, offered at a moment when the patient is oriented and able to respond, explaining the tool's role in documentation or decision support and giving the patient space to ask questions or decline it's usage. Where exigent circumstances make prospective disclosure impossible, disclosure should be made as soon as possible. Written agency policy should specify the disclosure language, documentation requirements, and any exceptions that may apply.

- **Clinician education and support:** Every EMS clinician using AI must be trained not just in functionality but also in ethics, accountability, and judgment. Reinforce the message: The tool assists—but you decide.
- **Workflow integration:** Do not add friction to field operations. AI tools must support clinicians, not distract them. Systems should be intuitive, mobile-ready, and responsive to the EMS pace.
- **Continuous monitoring and post-deployment QA:** AI is never "set and forget." Every tool must be monitored in real-world use with feedback loops for outcome tracking, user experience, and potential failure patterns.
- **Accountability and adverse event management:** Agencies must have protocols for documenting, investigating, and resolving AI-related errors. Accountability must extend across developers, agencies, and individual users.

- **Ongoing Considerations**
  - **Public engagement and trust:** EMS leaders must earn public trust. That means transparency, plain-language summaries of AI tools in use, and open communication about how AI supports patient care.
  - **Regulatory collaboration:** EMS agencies must work proactively with regulators at the state and federal levels to shape policies that reflect EMS realities and protect patient rights.
  - **Ethical innovation and research:** As technology evolves, so must our standards. EMS must remain engaged in AI research, ethics discussions, and the refinement of best practices.

Ethics Is Leadership in Practice

AI is not a shortcut. It is not a substitute for clinical responsibility or critical thinking. Every deployment of AI in EMS is a leadership decision that reflects our values, priorities, and commitment to the people we serve.

The ethical integration of AI begins with the decisions we make now. This framework is not exhaustive and is not final. But it provides the foundation for moving forward with integrity, accountability, and purpose.

If EMS waits for perfect consensus, we will inherit systems designed without us. If we delay action, we will be governed by technologies and policies that reflect someone else's vision, not ours. We must begin with what we know: professional judgment, clinical accountability, and public trust.

AI is not just another tool. It is a test of our leadership, ethics, and readiness to shape the future of emergency care.

Ethics is not the barrier to AI in EMS—it is the gateway, and leadership is the key.

Key Terms and Definitions

- **AI Risk Management Framework:** Guidelines from the NIST provide a structure for evaluating the safety, reliability, privacy, and ethics of AI systems across sectors, including health care.
- **Algorithmic Fairness:** The principle that AI systems should not produce biased or discriminatory outcomes based on race, gender, language, or other personal attributes. Requires deliberate model design and testing against diverse datasets.
- **Bias Amplification:** A phenomenon in which AI systems replicate and exacerbate existing disparities in training data, particularly when those disparities reflect historical inequities in healthcare delivery.
- **Clinician Oversight:** The requirement that licensed healthcare professionals retain final decision-making authority over AI-supported recommendations, ensuring human judgment remains central to patient care.
- **Data Governance:** The processes and policies that define how healthcare data is collected, validated, stored, protected, and used—particularly relevant in training ethical AI models.
- **Explainability:** A core principle of ethical AI design that ensures clinicians understand and communicate how and why an AI tool generated a specific output, score, or recommendation.
- **Federated Learning:** A privacy-preserving machine learning approach that trains AI models across decentralized data sources without transferring raw data to a central repository.

- **Health Equity Lens:** An evaluative framework to assess whether healthcare tools, systems, or decisions promote fair and just treatment for all populations, especially marginalized or historically underserved groups.
- **Human in the Loop:** A design model where clinicians actively engage with AI systems, reviewing, accepting, or overriding outputs based on context, training, and professional judgment.
- **NEMSIS (National EMS Information System):** The national database of EMS patient care records in the U.S., used for public health surveillance, research, and AI model development. Data quality and consistency in NEMSIS are foundational to ethical AI.
- **Privacy by Design:** A foundational privacy principle ensuring that data protection is embedded in system architecture and workflows from the outset, rather than as an afterthought.
- **Transparency:** The obligation of AI systems to disclose how they operate, what data they use, and what assumptions they rely on—critical for building trust in clinical settings.

Reflection Questions

1. Think about your current documentation practices. If the data you submit today helps train tomorrow's EMS AI systems, are you confident it reflects clinical reality, or just what fits in the boxes?
2. What systems does your agency have in place to review, validate, or improve data quality before it is shared externally or used in analytics? Who owns that responsibility?
3. When AI tools suggest a clinical course of action, how should clinicians evaluate and document their acceptance—or rejection—of that advice? What training is needed to support that?
4. How do we ensure equity in AI tools used in EMS? What questions should EMS leaders ask vendors, developers, and researchers before adoption?
5. Do your current QA (or quality improvement) processes include an ethical review of how AI is used in clinical decision-making? If not, what needs to change to add that oversight?
6. What does accountability look like when an AI tool fails—or succeeds? Who is responsible when an AI system misguides care, and how should agencies plan for that?

7. In your role, how will you advocate for AI tools that enhance, not replace, clinician judgment, protect patient dignity, and reflect the values of the EMS profession?

Study Questions

1. What real-world risks are introduced when AI systems in EMS are trained on incomplete or biased data?
2. Which of the following was a key finding from the 2019 *Science* study on healthcare algorithms?
   a. AI improved diagnostic accuracy across all populations
   b. Black patients were systematically deprioritized due to flawed cost-based proxies
   c. AI performed best when trained on international datasets
   d. Patients preferred AI over human clinicians for triage decisions
3. Define *explainability* in the context of EMS AI systems. Why is it essential for both patient trust and clinician accountability?
4. In your own words, explain the difference between algorithmic fairness and data accuracy, and why both are critical for ethical AI in EMS.
5. Ethics in EMS AI does not begin with the algorithm. It starts with __________.
6. According to the chapter, what is the primary ethical responsibility of EMS clinicians when using AI tools?
7. What core principle does the NIST AI Risk Management Framework emphasize for trustworthy AI systems?
   a. Commercial scalability
   b. Regulatory simplicity
   c. Validity, transparency, and human oversight
   d. Speed of implementation
8. What two additional pillars are included in the EMS-specific ethical AI framework that are not typically emphasized in hospital-based frameworks?
9. Why is narrative fidelity important in AI-generated EMS documentation?
10. List three actions EMS leaders or agencies must take to ensure ethical AI deployment in clinical practice.

CHAPTER SUMMARY

The Stakes Are Too High to Get This Wrong

AI is already reshaping medicine, and EMS is no exception. AI is rapidly entering the care environment, from triage prioritization to real-time decision support. But here's the imperative: If we trust AI to support patient care, we must build it with the same integrity we expect from our clinicians.

This chapter laid out a core truth: You cannot build ethical AI on unethical data. In EMS, that means improving how we document, securing how we store, and standardizing how we share. NEMSIS is one of the most powerful clinical datasets in health care, but it is only as trustworthy as the inputs it receives. We should not expect fairness, accuracy, or safety if we feed AI systems incomplete, biased, or inconsistent records.

We have already seen the consequences of flawed AI design. In the now-famous 2019 *Science* study, a healthcare algorithm underprioritized Black patients because it equated healthcare cost with clinical need. Bias built into the data became bias codified in care. EMS cannot afford to make that mistake at scale.

This is a call to action for individual clinicians, system leaders, educators, medical directors, vendors, and regulators. If we are embedding AI into our clinical workflows—and we should—we must do so transparently, equitably, and accountably. As the NIST AI Risk Management Framework clarifies, explainability, reliability, fairness, and human oversight are not idealistic—they are now foundational.

This chapter reaffirmed a critical standard: The clinician is still in charge. AI can support decisions, but it cannot carry responsibility. EMS clinicians must be trained to engage with AI—critically, thoughtfully, and confidently. That means knowing when to follow, when to question, and how to document the difference. Accountability does not end when the algorithm starts. It begins with us.

If EMS is to thrive in the era of intelligent tools, our systems must be high-tech and high-trust. That trust is not given; it must be earned through integrity in design, clarity in deployment, and leadership at every level of care.

12 Teaching Tomorrow

Preparing Emergency Medical Services Clinicians for Artificial Intelligence

LEARNING OBJECTIVES

After completing this chapter, the reader will be able to

- Describe the challenges EMS educators face in integrating AI into prehospital training, including concerns about cognitive overload and the risk of deskilling.
- Summarize the transformative potential of AI in EMS education, including its capacity to personalize learning, enhance diagnostics, and simulate complex scenarios.
- Compare EMS educational approaches with models adopted by other health professions, including nursing and medicine, and identify opportunities for interdisciplinary alignment.
- Explain how AI tools should be introduced in a way that maintains clinician authority, critical thinking, and human connection, even in high-tech environments.
- Apply the AAMC's AI Education Principles to EMS education, translating national best practices into actionable steps for EMS programs.
- Evaluate practical strategies for introducing AI literacy in EMS curricula, including case-based learning, structured debriefs, and critical thinking exercises.
- Discuss the ethical considerations related to AI use in EMS training, such as bias, data privacy, transparency, and the risk of overreliance.
- Identify key stakeholders, including regulators, EMS employers, academic partners, and interprofessional collaborators, needed to update EMS education for the AI era.

Prologue: A Veteran Educator's Reflection

It is difficult to admit, but our ambulances still carried LifePak 5s and 10s when I was a paramedic student. Two large-bore IVs wide open was standard for trauma, and a jaw screw was in the airway bag for seizure patients. Military antishock trousers were still in use, and 12-lead electrocardiograms had not yet entered the field.

When 12-lead electrocardiograms finally arrived, a few years into my career, they were considered revolutionary and burdensome. Adding one more assessment tool to an already overstuffed curriculum felt impossible. I later became a paramedic program director at a community college. Around that time, intraosseous access and King airways were introduced. Once again, we faced the same pressure: too many skills and not enough hours.

Fast forward to today. The number of required hours for emergency medical technician and paramedic education has barely changed. But what students are expected to learn has exploded. Continuous positive airway pressure, bilevel positive airway pressure, whole blood administration, advanced clinical algorithms, expanded pharmacology, electronic charting, EMS billing—every piece is more complex than the last.

So, when I raise the idea of integrating AI into EMS education, I am not surprised by the looks I get from other educators. Some react with skepticism, others with visible frustration. Their concerns are valid. They worry about cognitive overload. At the same time, they fear that students will use AI as a shortcut—an excuse not to master clinical reasoning or documentation.

They are right to be concerned. These challenges are real—but the tools are coming, whether we are ready or not. Our job now is to ensure that EMS education does not just react to the future. It must prepare clinicians to lead us through it.

Reimagining Emergency Medical Services Education for the Age of Artificial Intelligence

Resistance to AI in EMS education is real. Educators consistently raise two concerns.

1. **Educational overload**

   EMS instructors are asked to do more with fixed resources. While other health professions use flexible, competency-based education, EMS remains tied to outdated hour-based requirements. The knowledge base has expanded, but the hours have not. Instructors carry the weight of this imbalance, which is the unfortunate reality of the environment in which EMS educators operate.

2. **Fear of deskilling**

   There is also a concern that AI will erode critical thinking. If clinicians rely on automation, foundational judgment may weaken. This is a legitimate risk which is why AI must not replace human decision-making; it should enhance it. When properly used, AI sharpens awareness, adds depth to differential thinking, and strengthens documentation. This way the clinician remains in command.

The Transformative Potential of Artificial Intelligence in Emergency Medical Services

In 2023, a journal publication explored the transformative potential of AI technologies, such as OpenAI's ChatGPT, within EMS systems and education. The authors point out several key benefits of AI, including its ability to enhance diagnostic capabilities through natural language processing, create immersive training simulations, and offer personalized learning pathways for EMS clinicians.[1]

These technologies hold substantial promise. AI can be leveraged to improve diagnostic accuracy by analyzing vast amounts of data in real-time and identifying trends that human clinicians might miss. It can create realistic, scenario-based training environments that allow EMS students to respond to complex cases in a controlled setting. Most importantly, AI can provide individualized learning experiences, adapting to the clinician's current level of competence and guiding them toward mastery.

Lessons from the Field: The Unique Reality of Emergency Medical Services

EMS operates under unique, high-pressure conditions. When an emergency medical technician or paramedic treats a critically ill patient, they often work in a tiny team, sometimes just one or two people. In contrast, a patient in a hospital setting might be cared for by a team of five, ten, or even more professionals. This resource difference underscores the need for EMS clinicians to maximize every available tool.

In the field, AI is not merely a convenience. It is essential. AI can act as an invaluable partner, offering a second opinion, suggesting differential diagnoses, and ensuring that documentation is accurate and complete. EMS clinicians often make life-or-death decisions with limited support, and AI can ensure they have the best possible information at their fingertips.

For EMS professionals, embracing AI does not replace human expertise but *augments* it. Just as paramedics use advanced equipment and drugs to enhance patient care, AI should be seen as a tool to improve clinical outcomes and streamline decision-making.

Reclaiming the Human Core of Emergency Medical Services

Despite its incredible capabilities, AI will never replace the human connection at the heart of EMS. While AI will excel in tasks such as generating differential diagnoses or submitting protocols instantly, it cannot provide the emotional support often required in critical moments.

AI cannot hold a scared child's hand during a seizure or offer a calming presence to a combative trauma victim. These are the moments that define the very essence of our profession. Empathy, compassion, and connection are irreplaceable.

Any integration of technology must be balanced with humanity. EMS clinicians must prioritize human connection even as they embrace AI's technological advancements. The ability to recognize the person behind the patient data remains the core of EMS practice, and this balance of heart and technology will define the future of emergency medical care.

Empathy, compassion, and connection are irreplaceable.

Lessons From Medicine: What Emergency Medical Services Can Learn from Other Professions

Across health care, educators no longer debate whether AI belongs in the classroom. They are defining *how* to integrate it—ethically, practically, and with urgency.

Medical schools, nursing programs, and hospital systems have already begun preparing their students for AI-enabled care. The American Medical Association, the Association of American Medical Colleges (AAMC) and academic leaders at University of California San Francisco and the University of Michigan are piloting curricula that embed AI into clinical reasoning, diagnostics, and decision support.

At University of California San Francisco, Dr. Cornelius James emphasizes that AI should be introduced not as a replacement but as a diagnostic teammate. His model trains students to assess AI strengths and weaknesses using real-time structured reflection and calibration methods. It reinforces the core values of medicine—humility, empathy, and human connection—even in the presence of high-tech tools.

The American Nurses Association highlights the need for continuous education and professional development. Their position is clear: Nurses must be trained to work effectively alongside AI. This includes building AI literacy into prelicensure nursing education and ongoing career development.

The American College of Physicians takes this further. Their policy framework mandates comprehensive education at all levels of training, ensuring physicians are equipped to practice in AI-enabled environments. They stress the importance of unified regulatory oversight to ensure safe, ethical implementation.

Leading medical institutions are operationalizing these principles. Harvard Medical School, for example, now offers an AI-in-medicine doctoral track. Dean Bernard Chang puts it plainly:

> *I do not believe that AI will replace human physicians, but those who use AI will be so much more capable than those who do not.*[2]

Harvard's focus is foundational: Students are learning AI as part of how medicine works, not as an add-on skill. Duke University School of Medicine is also taking deliberate steps to align AI education with its clinical and research priorities. According to Dr. Aditee Narayan, AI instruction must be directly tied to why people enter medicine in the first place: to improve health outcomes through service, care, and innovation.[3]

This is the turning point EMS must recognize. EMS clinicians deserve the same educational investment and institutional support as their hospital-based peers. Training EMS clinicians to engage with AI must include the following:

- Interpreting and contextualizing AI-generated insights
- Understanding algorithmic limitations and bias
- Protecting patient privacy and data integrity
- Maintaining ethical boundaries
- Overriding AI recommendations when necessary

EMS educators should not view AI as a technical feature. It is a clinical reality. Students must learn to apply it with judgment, skepticism, and care. That means rethinking how we teach documentation, triage, pharmacology, and differential diagnosis—all of which machine intelligence will soon influence.

EMS organizations and educational institutions must collaborate to develop training and certification programs that reflect this reality. The urgency cannot be overstated. AI education should have been a priority years ago. It is now nonnegotiable.

As technology advances, so must our training. The EMS profession must lead, not follow, on this front.

Integrating Artificial Intelligence Literacy into Emergency Medical Services Education

The good news is that EMS educators do not have to build from scratch. We already teach critical thinking, systems awareness, and patient-centered care. AI literacy can—and should—be layered into these existing structures.

Here is how to start:

- **Foundation building:** Introduce brief modules or guest lectures on key AI concepts, such as machine learning, large language models, and clinical applications of natural language processing.
- **Scenario-based learning:** Incorporate AI tools into case studies and let students evaluate AI-generated differentials, flag errors, and defend their own decisions.

- **Critical thinking challenges:** Present flawed AI outputs intentionally and require students to identify limitations, correct recommendations, and explain their reasoning.
- **Human factors debriefs:** Use post-simulation discussions to reinforce empathy, cultural awareness, and communication skills that AI cannot replicate.

These methods require no major overhaul—just intentional integration. They prepare clinicians to lead with discernment, not dependency.

Adapting the Association of American Medical Colleges Artificial Intelligence Principles for Emergency Medical Services Education

To guide the ethical, effective integration of AI into EMS education, the AAMC[4] has established a set of principles that can be directly adapted for EMS. These are not abstract ideals—they are practical guardrails designed to ensure AI supports clinicians without undermining their judgment, equity, or ethical responsibility. EMS educators must lead in applying these principles across didactic content, simulation training, and clinical evaluation. Table 12–1 outlines how each principle translates into actionable guidance for EMS instruction.

These principles can be integrated into didactic and scenario-based learning to ensure AI is introduced as a tool of responsibility, not just efficiency.[5]

Reclaiming the Human Core

AI will eventually outperform clinicians on many technical tasks—such as diagnosis, documentation, and protocol adherence—but it will never replace human connection.

What EMS needs now is not less professionalism but more. Our students must learn to

- hold a scared child's hand during a seizure,
- speak gently with a combative trauma victim, and
- see the person behind the data.

By integrating AI into education, we do not dilute these values. We make space for them to shine.

TABLE 12–1. AAMC Principles Applied to EMS

Principle	Application
1. Human-Centered	AI must enhance—not replace—the judgment, compassion, and professionalism of EMS clinicians.
2. Equity-Focused	Educators must vet AI tools for bias and ensure they do not reinforce disparities in care.
3. Transparent and Explainable	Clinicians must understand how AI reaches conclusions. Black-box systems are unacceptable in EMS.
4. Interdisciplinary Collaboration	EMS educators should partner with AI developers, physicians, nurses, and regulators to ensure aligned training.
5. Continuous Learning	AI use must evolve with clinical feedback and reflect real-world EMS practice.
6. Privacy and Data Stewardship	EMS training must emphasize protection of patient data and ethical handling of digital records.
7. Ethical Implementation	Teach students not just what AI can do—but what it *should* do. Clear ethical standards are essential.

Seizing the Leadership Opportunity

AI is no longer on the horizon—it is already reshaping emergency care. EMS educators, clinicians, and leaders must now ensure that our workforce is prepared technically, ethically, and professionally.

We have clarified that EMS will not be replaced by AI. But those who fail to adapt will be outpaced—clinically, operationally, and professionally—by those who do.

The debate over whether AI belongs in EMS education is over. What matters now is *how* we integrate it—and whether agencies, accreditors, and stakeholders will support the structural changes needed to make it possible. That includes revisiting rigid hour-based models that no longer reflect the scope and complexity of modern EMS practice.

This is a defining moment for our profession. We can lead the way, building a future where technology strengthens care, supports clinicians, and protects what makes EMS uniquely human. Alternatively, we can hesitate and allow that future to be built without us.

Leadership requires action and change—and the time to act is now.

Key Terms and Definitions

- **AI Literacy:** The ability to understand, interpret, and critically evaluate AI tools and outputs within clinical and educational contexts.
- **Algorithmic Bias:** Systematic and unfair outcomes produced by AI due to skewed or unrepresentative training data, often reinforcing existing healthcare disparities.
- **Competency-Based Education:** A model of instruction that emphasizes demonstrated skills and understanding over time-based instruction, allowing learners to progress at their own pace and depth.
- **Continuous Learning:** An educational approach emphasizing ongoing, lifelong adaptation of skills and knowledge, especially important as AI tools evolve.
- **Deskilling:** A potential risk in technology-enhanced education, where overreliance on automation may diminish foundational clinical reasoning and hands-on skills.
- **Equity-Focused Education:** Instructional design that proactively addresses disparities in access, representation, and outcomes, especially relevant when introducing AI tools trained on biased datasets.
- **Explainability:** The degree to which clinicians and students can understand the rationale behind an AI recommendation, score, or output. Essential for safe, ethical use of AI in education and practice.
- **Human-Centered Design:** An approach to technology and education that keeps the human experience—empathy, professionalism, and patient connection—at the core of all instructional or clinical decisions.
- **Natural Language Processing:** A branch of AI focused on understanding and generating human language. Used in EMS education for generating differentials, documentation assistance, and voice-to-chart systems.
- **Scenario-Based Learning:** An educational method using real-world clinical simulations or cases to reinforce knowledge, judgment, and application of tools like AI.
- **Technology-Enhanced Learning:** The integration of digital tools—including AI—into the classroom or field education to support knowledge acquisition, skill-building, and critical thinking.

Reflection Questions

1. What specific changes would your EMS program need to make to integrate AI literacy into the existing curriculum? Consider scope,

faculty training, and whether current hour requirements allow for meaningful adaptation.

2. How can EMS educators ensure that AI is seen as a clinical tool, not a shortcut, especially among students prone to over-relying on technology? What teaching strategies reinforce judgment, skepticism, and accountability?
3. Where is the balance between standardization and innovation in EMS education? How can we adopt national AI frameworks while preserving the flexibility to reflect local system needs?
4. What ethical concerns must be addressed before introducing AI tools into EMS simulation labs or clinical decision support training? Think in terms of patient privacy, bias, and clinician autonomy.
5. Who needs to be at the table to update the EMS educational model for the AI era? Consider partnerships with regulators, employers, software developers, and interprofessional allies.

Study Questions

1. What historical challenges have EMS educators faced when integrating new clinical tools or protocols into prehospital education?
2. What are two primary concerns EMS educators express about introducing AI into initial education and continuing professional development?
3. Which of the following is *not* listed as a benefit of AI integration in EMS education?

    a. Enhancing diagnostic reasoning
    b. Automating credentialing exams
    c. Creating immersive training simulations
    d. Offering personalized learning pathways

4. How does the field environment of EMS uniquely position AI as a clinical support tool, compared to hospital-based practice?
5. True / False: AI can and should replace empathy and human connection in EMS practice as it becomes more capable in diagnosis and documentation.
6. What are three key skills or competencies EMS students must develop to use AI effectively and ethically in clinical settings?
7. According to the chapter, how are leading medical schools (e.g., Harvard, Duke, UCSF) integrating AI into healthcare education?
8. The term _______ refers to the ability to understand, interpret, and critically evaluate AI tools and outputs within clinical and educational contexts.

9. According to the AAMC principles adapted for EMS education, which of the following is *essential* for ethical AI integration?

   a. Gamification of patient care
   b. Instructor-only access to AI
   c. Transparent and explainable algorithms
   d. Use of AI as a grading tool

10. Why is it critical that EMS education adapt to the presence of AI now, rather than wait for regulatory mandates or textbook revisions?

CHAPTER SUMMARY

Teaching Tomorrow—Preparing Emergency Medical Services Clinicians for Artificial Intelligence

EMS education is at a crossroads. AI is no longer theoretical—it is operational, evolving, and increasingly embedded in health care. Yet, EMS classrooms remain constrained by outdated hour-based models, overburdened instructors, and a curriculum that has not kept pace with the realities of modern prehospital care.

This chapter examined the barriers and opportunities that AI presents to EMS education. We addressed the valid concerns of overload and deskilling but also challenged educators to move beyond resistance and recognize the potential of AI to enhance diagnostic reasoning, personalize learning, and strengthen documentation.

Drawing lessons from medical schools, nursing programs, and national policy organizations, we outlined a path for EMS to lead, not follow, the integration of AI into clinical education. Tools like AI-powered simulations, natural language models, and predictive analytics are already in play. Our role is to ensure they are used with ethical oversight, clear educational standards, and unwavering commitment to human-centered care.

AI will change how we teach, but it will not change why we teach. The next generation of EMS clinicians must be prepared to think critically, act ethically, and lead confidently in a world where human skill and machine intelligence work side by side.

13

The Strategic Integration of Artificial Intelligence in Emergency Medical Services

LEARNING OBJECTIVES

After completing this chapter, the reader will be able to

- Analyze the critical challenges facing EMS systems, including workforce shortages, financial instability, and escalating call complexity.
- Describe the potential role of AI in enhancing EMS sustainability, including automation, optimized staffing, and data-driven reimbursement.
- Evaluate how AI tools can reduce clinician burden, improve recruitment and retention, and support real-time decision-making during high-acuity events.
- Discuss the application of AI to improve patient care, including early warning systems, diagnostic support, and personalized intervention pathways.
- Identify opportunities for AI to enhance operational efficiency, such as predictive dispatch, route optimization, and system performance monitoring.
- Explain how continuous data analysis supported by AI can drive system improvement, support education, and foster adaptive, learning-focused EMS environments.
- Assess the ethical implications of AI integration, including privacy, bias mitigation, and equitable care delivery in diverse communities.
- Formulate a strategic roadmap for AI integration, outlining the necessary investments, partnerships, regulatory considerations, and leadership actions required for success.

Across the nation, EMS systems are facing a crisis of unprecedented proportions, grappling with multifaceted challenges that threaten their very foundation. Financial sustainability remains a persistent issue, with limited funding and escalating operational costs creating a precarious fiscal environment. At the same time, workforce recruitment and retention have reached critical levels, with EMS agencies struggling to attract and keep skilled professionals. The demands on EMS services continue to escalate, driven by increasing call volumes, more complex medical emergencies, and the expectation of rapid response times.

EMS clinicians are at the frontline of this struggle, enduring high stress, long hours, and an ever-growing operational burden. These dedicated professionals are often stretched to their limits, leading to severe burnout and alarming turnover rates. The departure of experienced personnel not only diminishes the quality of care but also places additional strain on those who remain, creating a vicious cycle of overwork and attrition.

The compounded effect of these challenges threatens EMS systems' reliability, efficiency, and effectiveness. As the backbone of emergency medical response, any compromise in their functionality can have dire consequences for community health and safety. The growing complexity of health care further exacerbates the situation, which requires EMS providers to adapt quickly and efficiently.

The situation in Colorado exemplifies the broader national crisis. According to the Colorado Emergency Medical Services System, *Sustainability Task Force Phase I Report*, the state's EMS systems are under significant strain due to workforce shortages and financial instability.[1] In many regions, EMS providers report difficulty recruiting new staff and retaining current employees, with turnover rates exceeding 20% annually in some areas. Financial challenges are equally severe, with many EMS agencies operating at a deficit due to insufficient reimbursement rates and rising operational costs. These economic constraints limit the ability to invest in essential equipment, training, and technology, further exacerbating the workforce crisis and hindering the delivery of high-quality care.

The Opportunity for Modernization

While AI is not a magic solution, it presents a new opportunity for EMS to modernize and leverage technology to address these pressing issues. Much like other sectors of health care have realized the transformative potential of AI, EMS has the chance to adopt advanced technologies to enhance efficiency, improve patient outcomes, and create a more sustainable and resilient system.

Enhancing Financial Sustainability

AI can play a crucial role in improving the financial sustainability of EMS systems. AI can reduce costs and increase efficiency by optimizing resource allocation and streamlining operations. Predictive analytics can help forecast demand, allowing EMS organizations to manage their resources better and reduce unnecessary expenditures. Furthermore, AI-driven billing and coding systems can enhance revenue cycles by ensuring accurate documentation and maximizing reimbursement rates.

Automating routine administrative tasks with AI can also free up valuable time for EMS clinicians, allowing them to focus on patient care rather than paperwork. This shift improves job satisfaction and enhances operational efficiency, contributing to the overall financial health of EMS organizations.

Addressing Workforce Recruitment and Retention

Workforce recruitment and retention are critical challenges for EMS. High turnover rates and difficulties attracting new talent exacerbate the strain on existing staff. AI can help alleviate these issues by reducing the cognitive load and administrative burden on EMS clinicians. Real-time decision support systems and AI-powered documentation tools streamline workflows, enabling practitioners to deliver care more efficiently and effectively.

Moreover, AI can enhance training and professional development for EMS personnel. AI-powered simulation-based training programs provide realistic, data-driven scenarios that help practitioners hone their skills and prepare for emergencies. Continuous learning and improvement through AI-driven feedback can boost job satisfaction and retention as EMS clinicians feel more competent and supported.

Improving Patient Outcomes

The integration of AI into EMS offers unparalleled opportunities to improve patient outcomes. Real-time decision support, predictive analytics, and advanced diagnostic tools enable EMS clinicians to make more informed and timely decisions. AI can quickly analyze vast amounts of data, identifying patterns and trends that human practitioners might miss. This capability is crucial in high-pressure situations where every second counts.

AI-powered tools can provide early warnings about patient deterioration, suggest optimal treatment protocols, and predict the likelihood of certain medical conditions based on real-time data analysis. By incorporating AI into their practice, EMS teams can deliver more precise and effective care, ultimately improving patient survival and recovery times.

Operational Efficiency and Resource Management

Limited resources and high demand often strain EMS systems. AI can be pivotal in optimizing resource allocation, streamlining operations, and reducing response times. Through predictive modeling and real-time analytics, AI can forecast service demand, identify peak periods, and optimize the deployment of ambulances and medical personnel. This proactive approach can significantly enhance the efficiency of EMS operations, ensuring that resources are available where and when needed.

Additionally, AI can assist in route optimization, helping EMS teams navigate traffic and other obstacles more effectively. This can lead to faster response times and better patient outcomes. By automating routine administrative tasks, AI can also free up valuable time for EMS clinicians, allowing them to focus more on patient care and less on paperwork.

Data-Driven Insights and Continuous Improvement

The integration of AI into EMS enables the collection and analysis of vast amounts of data, providing valuable insights into the effectiveness of medical interventions and overall system performance. By continuously analyzing this data, EMS organizations can identify areas for improvement, refine their protocols, and implement best practices. This data-driven approach fosters a culture of continuous improvement, ensuring that EMS systems evolve and adapt to changing needs and emerging challenges.

AI can also facilitate training and education by providing EMS clinicians with feedback and insights based on real-world data. Simulation-based training programs powered by AI can help practitioners hone their skills and prepare for a wide range of scenarios, ultimately leading to better preparedness and higher standards of care.

Ethical and Equitable Care

AI has the potential to promote more equitable care by reducing human biases and ensuring consistent application of medical protocols. By standardizing decision-making processes and providing objective data-driven recommendations, AI can help ensure that all patients receive the same high level of care, regardless of their background or circumstances. This is particularly important in diverse, underserved communities with more prevalent healthcare disparities.

The Strategic Vision for Artificial Intelligence Integration

The strategic integration of AI in EMS requires a clear vision and a deliberate approach. EMS organizations must prioritize the development of robust AI infrastructure, invest in training and education, and foster a culture of innovation and adaptability. Collaboration with technology developers, policymakers, and other stakeholders is essential to create an ecosystem that supports the effective implementation of AI.

Moreover, the integration of AI should be guided by ethical principles and a commitment to patient privacy and data security. Ensuring that AI systems comply with regulatory standards and protecting patient information are critical components of a successful AI strategy. Integrating AI into EMS is a technological upgrade and a strategic imperative that can revolutionize patient care, enhance operational efficiency, and drive continuous improvement. By embracing AI, EMS organizations can position themselves at the forefront of medical innovation, delivering higher standards of care and better patient outcomes. The future of EMS lies in the seamless integration of AI, and EMS leaders must recognize and act on this transformative potential.

Key Terms and Definitions

- **AI Infrastructure:** The combination of hardware, software, data systems, and operational protocols needed to support the deployment and maintenance of AI tools in EMS environments.
- **Administrative Automation:** The use of AI tools to complete or support nonclinical tasks such as billing, coding, documentation, and resource scheduling, thereby reducing clinician workload.
- **Cognitive Offloading:** A process where AI systems reduce mental strain on clinicians by assisting with decision-making, calculations, or pattern recognition, improving safety and focus under stress.
- **Continuous Improvement:** An iterative approach to operational excellence, where EMS systems use data analytics to refine protocols, improve care delivery, and evolve workforce practices.
- **Data-Driven Decision-Making:** Using real-time or retrospective analytics to guide EMS systems' clinical, operational, or strategic choices.

- **Decision Support Systems:** AI-based tools that assist EMS clinicians by analyzing patient data and recommending potential diagnoses, interventions, or triage priorities.
- **Equitable Care:** Healthcare delivery that ensures fair and impartial treatment of all patients, regardless of race, income, geography, or other social determinants—supported by unbiased AI systems.
- **Operational Efficiency:** The ability to deliver timely, high-quality EMS services with optimized use of staff, vehicles, and other resources.
- **Predictive Analytics:** AI-powered forecasting methods that identify future EMS needs, such as call volume, resource demand, or patient risk, based on historical and real-time data.
- **Strategic Integration:** A long-term, coordinated approach to adopting AI in EMS that aligns with organizational goals, workforce needs, clinical priorities, and public health imperatives.

Reflection Questions

1. Which aspect of your EMS system—workforce stability, finances, clinical accuracy, or operational efficiency—stands to benefit most from AI integration? Why?
2. What current administrative tasks in your agency could be automated with AI, freeing up time and energy for clinical care?
3. What ethical or regulatory safeguards must be in place before your agency adopts AI tools for clinical support or dispatch optimization?
4. How can EMS systems ensure that AI enhances—not replaces—clinicians' expertise, autonomy, and empathy?
5. If you were tasked with leading AI integration in your EMS agency, who would need to be at the table, and what would be your first three priorities?
6. How can your agency measure the impact of AI tools on patient outcomes, workforce well-being, and financial performance?

Study Questions

1. What three core challenges are currently threatening the stability and reliability of EMS systems across the United States?

2. Which of the following best describes how AI can improve financial sustainability in EMS?

   a. By eliminating the need for emergency medical technician certification
   b. By replacing billing departments with outsourcing services
   c. By optimizing resource allocation, automating documentation, and improving reimbursement
   d. By automating emergency dispatch centers completely

3. How can AI tools contribute to better recruitment and retention of EMS clinicians?
4. True / False: AI decision support systems are intended to replace the EMS clinician's judgment in high-acuity situations.
5. What is the primary way predictive analytics can improve EMS operational efficiency?

   a. By generating automated social media alerts
   b. By forecasting demand, optimizing deployment, and reducing response times
   c. By teaching new clinical protocols
   d. By calculating annual salary budgets for EMS agencies

6. Name two ways AI can support continuous improvement in EMS systems.
7. In what ways does AI integration promote more equitable care across diverse communities?
8. Which of the following is a core requirement for successful AI integration in EMS?

   a. Permanent suspension of quality assurance reviews
   b. Eliminating medical direction for low-acuity calls
   c. Clear leadership, ethical governance, and investment in AI infrastructure
   d. Outsourcing field care to remote providers

9. AI systems that reduce mental strain on clinicians by assisting with decisions, calculations, or pattern recognition are said to provide _______________.
10. What are three key components of a strategic roadmap for integrating AI into EMS practice?

CHAPTER SUMMARY

The Strategic Integration of Artificial Intelligence in Emergency Medical Services

Chapter 13 addresses the urgent challenges facing EMS and articulates a strategic case for integrating AI as a transformative solution. EMS systems across the United States are confronting significant pressures—financial constraints, escalating demand, and workforce challenges—but these same forces present an opportunity to reimagine how care is delivered and supported. These pressures threaten the quality, timeliness, and reliability of prehospital care.

In response, this chapter frames AI not as a cure-all, but as a strategic enabler—capable of enhancing system resilience, clinician performance, and patient outcomes. By leveraging AI-powered tools, EMS can optimize resource allocation, streamline administrative tasks, and automate documentation processes, improving financial efficiency and freeing up clinicians for patient-centered care. Real-time decision support and predictive analytics can augment clinical accuracy, support earlier interventions, and enable data-driven triage, particularly in high-stress or under-resourced settings.

The chapter further explores how AI can support clinician recruitment, reduce burnout through cognitive offloading, and elevate training via simulation-based learning. It emphasizes that strategic AI adoption must be grounded in ethics, equity, and regulatory compliance. A clear, vision-aligned integration plan—anchored in transparency, interdisciplinary collaboration, and clinician leadership—is essential. If implemented intentionally, AI offers EMS a path toward modernization, sustainability, and professional transformation.

14

Embracing the Future of Emergency Medical Services with Artificial Intelligence

LEARNING OBJECTIVES

After reading this chapter, the reader will be able to

- Describe the historical significance of EMS's recognition as an allied health profession, and how that history informs today's leadership challenge around AI.
- Explain why AI integration in EMS is an urgent and strategic necessity, not a theoretical or optional development.
- Identify the key roles EMS leaders, educators, and clinicians must play in ensuring ethical, effective, and responsible AI adoption.
- Compare AI integration efforts in other healthcare professions and assess what lessons EMS can draw from those examples.
- Articulate the risks of EMS fragmentation or inaction, and describe the importance of a unified, standards-based approach to technology adoption.
- Define core leadership traits necessary to guide EMS through the AI era, including curiosity, collaboration, adaptability, and a commitment to equity and excellence.
- Assess how consistent national standards can strengthen EMS's role in healthcare innovation and help ensure equitable access to AI-powered tools across diverse systems.

In 1975, Dr. J.D. "Deke" Farrington—a respected orthopedic surgeon and Chair of the National Registry of Emergency Medical Technicians—wrote to the American Medical Association with a bold request: that EMTs and paramedics be formally recognized as an allied health profession. The American Medical Association agreed. This decision marked a pivotal moment in EMS history. It confirmed that paramedicine, when unified by national certification, standards, and accreditation, belonged alongside medicine, nursing, and other respected healthcare professions.

EMS did not earn that recognition by default. It earned it through clarity of purpose, agency cohesion, and a shared vision for what the profession could become.

Today, EMS faces another defining moment—not one of identity, but of trajectory.

This time, the opportunity is not a single policy or certification—it is a convergence of tools, systems, and technologies that have the potential to transform how EMS delivers care, coordinates operations, and protects communities.

AI is part of that future, as are smart documentation platforms, wearable monitors, predictive dispatch systems, virtual care models, language translation apps, and autonomous transport innovations. Technology alone will not define the next era of EMS, but how we choose to lead it will.

These tools offer more than automation. With new capabilities, they offer a way to reimagine old constraints—workforce shortages, time pressure, fragmented data, and documentation burdens. When combined with courageous leadership, intelligent systems can potentially reposition EMS to survive change and shape it, not just to catch up to healthcare transformation but to help lead it.

But this will not happen by accident. It will require intention, alignment, and a shared commitment to the values that have carried EMS for decades.

What This Book Has Shown

Throughout this book, we have explored both the promise and the pitfalls of emerging technologies:

- We have seen how AI can support—not replace—clinical decision-making.
- We have highlighted the enduring importance of the EMS narrative as both a legal record and a human story.
- We have named the cultural and cognitive biases that often resist innovation and offered ways to move past them.

- We have learned that equity must be a requirement, not a byproduct, of digital health tools.
- We have examined national policy momentum, healthcare investment trends, and global pilots that position EMS at a critical crossroads.
- And we have returned, again and again, to the core truth: Technology does not lead. *People do.*

Health Care or Transport to Health Care? The Decision Each Agency Will Now Make

In 1975, Dr. Farrington made the case for EMS to the American Medical Association as a new medical profession. The AMA agreed because it was a decision made at the profession level, by professional-level institutions, about whether EMS belonged alongside nursing, medicine, and the allied health disciplines.

The Decision This Era Forces on Emergency Medical Services Is Different in Kind

The question of whether EMS is health care itself or just transport to health care has been asked in our profession for decades, and I have spent much of my career arguing the former. What has changed is my view on who gets to answer. I no longer believe the profession will answer this question as a profession. I believe each agency will answer for itself, and that the two answers will both become real. Some EMS agencies will integrate deeply into the healthcare continuum: interoperable documentation, shared records, ambient AI documentation, clinician-led AI governance, seat at the regional health information exchange table, participation in community paramedicine and population health, and a strong embrace of education and university degrees. Other agencies will define themselves as transport: essential, professional, high-quality emergency response and conveyance, and will leave the integrated healthcare work to the agencies that choose it. Both paths will exist. Both will be defensible. Both will carry consequences.

The healthcare path is harder. It requires investment in interoperability, real engagement with national policy, a higher standard of documentation, and a willingness to be held accountable against outcomes that were never part of the transport-era EMS contract. Agencies that walk this path will gain access to funding streams, clinical partnerships, and technological infrastructure that the transport-centric EMS agencies will not. They will also inherit obligations, disclosure to patients about AI use, data-sharing responsibilities, participation in quality measurement programs, that look more like hospital compliance than traditional EMS regulation.

The transport-to-healthcare EMS path is simpler. It is also narrower. Agencies on that path will continue to deliver the lifesaving work EMS has always done, and patients will continue to depend on them. But those agencies will increasingly find themselves operating within a healthcare system that routes decisions, records, and reimbursement around them rather than through them. The financial pressure that already threatens EMS sustainability in rural and underserved communities will intensify for agencies whose role in the healthcare system stops at the emergency department doors. Autonomous transport platforms, AI-mediated patient navigation tools, and consumer health apps do not care about an EMS agency's identity. They simply capture whatever workflow is willing to be captured.

I do not write this to argue one path is right for every agency. A small volunteer department in a frontier county has a legitimately different calculus than a large urban third-service department with a medical director and a data analyst on staff. The evolution of my own view is precisely this: I once argued the profession must choose. I now argue that the profession will not choose as a single body, and that this is fine, provided each agency chooses deliberately, with eyes open, and with leadership honest enough to tell its clinicians and its community which path it is walking and why.

The agencies choosing the healthcare path will define what AI-enabled EMS looks like in this country. They will write the disclosure policies. They will negotiate the data-sharing agreements. They will run the pilots that become the templates. Over time, they will constitute the critical mass that the profession was unable to assemble at the national level: not a unified voice, but a coherent one, earned through operational example rather than conferred by consensus.

The remainder of this chapter is directed to those agencies and to the clinicians and leaders inside them. If you are choosing the healthcare path, or you are inside an agency that has quietly begun walking it without naming the choice, what follows is for you.

What Emergency Medical Services Must Do Next

As EMS clinicians, educators, and leaders, we must ensure that innovation reflects our profession—not reshapes it without us. We must remain agile, collaborative, and unified—especially as national systems move toward real-time data, performance metrics, and technology-enabled care.

We do not need to chase every tool, but we do need to understand the direction and lead with values.

The future of EMS will belong to those who can do the following:

- Build bridges across technology and clinical practice
- Preserve judgment, ethics, and context in a digital world
- Advocate for smart regulation and national alignment
- Train a workforce that is curious, confident, and capable
- See every new tool as a test of leadership, not just a test of code

EMS does not need billions in venture capital to thrive. We need vision. We need shared purpose. We need the courage to act—together—before consensus arrives too late.

Principles to Carry Forward

If the first generation of EMS leaders built a profession, the next generation must build a future. At the foundation, our values are to

- **Embrace curiosity:** Seek knowledge. Explore tools. Anticipate the next challenge.
- **Work in collaboration:** Share ideas. Align across systems. Lead beyond silos.
- **Demand excellence:** Innovation must never excuse inaccuracy or inequity.
- **Value standards:** Consistency enables quality, safety, and trust.
- **Adapt to change:** Rigidity is the enemy of relevance.
- **Foster inclusivity:** Diverse voices are essential to ethical and practical design.
- **Innovate continuously:** The pace of technology will not slow, so neither can we.

These are not ideals. They are imperatives. They are the shared practices defining EMS leadership in a changing world.

A Night Call That Still Matters

It was after 3:00 a.m. when the call came in—"elderly female, difficulty breathing." A quiet block. A dim porch light. She met us at the door, walking but weary. No family around. No clear cause. Her vitals were borderline, her voice soft. "I just don't feel right," she said.

There was nothing that triggered an alert—no glaring abnormality, no clear protocol path—but the medic paused. He had been on this kind of call before, and he knew that sometimes the sickest patients do not shout—they whisper.

She was pale. The air felt heavy. Something was wrong, even if the monitor said otherwise.

They transported. She crashed fifteen minutes later en route. Rapid intervention saved her life.

No AI model would have flagged that moment today. No algorithm could have replicated the years of instinct, the subtle cues, or the human connection that led to that decision.

But one day—if we build wisely—it might.

If we embed empathy into design, translate judgment into code, and protect the clinician's voice at every stage of care, AI and emerging technologies will not just follow EMS but elevate it.

This book has never been about technology alone; it has been about *trust*, *responsibility*, and *leadership*. It is about the people who answer the call, at 3:00 a.m. or in the boardroom, and believe that care still begins not with a code, but with a choice.

The future of EMS is not something we await. It is something we must build *together*.

> *The best way to predict your future is to create it.*
>
> —President Abraham Lincoln

Key Terms and Definitions

- **AI Integration:** Incorporating AI technologies into EMS systems, including clinical decision support, documentation, dispatch, and operational management.
- **Allied Health Profession:** A classification for healthcare professions (such as EMS, nursing, and physical therapy) that deliver essential services but are not part of medicine, dentistry, or nursing. EMS gained this designation in part through national certification and professional standards.

- **Collaboration:** An interdisciplinary and interagency approach to integrating AI and innovation in EMS, ensuring unified strategies and broad stakeholder engagement.
- **Continuous Innovation:** An ongoing commitment to testing, adopting, and refining new technologies and methods that improve EMS delivery.
- **EMS Leadership:** Clinicians, educators, administrators, and policymakers responsible for guiding the vision, strategy, and ethical implementation of innovations, including AI, within EMS systems.
- **Ethical Implementation:** The integration of technology in a way that protects patient dignity, ensures data privacy, mitigates bias, and preserves human oversight.
- **Professional Standards:** Agreed-upon norms for education, certification, conduct, and operations that help define EMS as a unified, respected health profession.
- **Strategic Vision:** A long-term, values-driven roadmap for how EMS will adopt AI to meet clinical, operational, and workforce goals without losing its foundational mission.
- **Unified Voice:** Collective action and shared policy direction across EMS agencies, leaders, and educators that strengthens the profession's influence and cohesion.

Reflection Questions

1. What risks do you see if EMS delays AI adoption compared to other healthcare professions? How might that impact the profession's long-term credibility or relevance?
2. Who in your organization or state is best positioned to lead AI integration efforts, and what support or education do they need to succeed?
3. How can your agency ensure that AI strengthens professional standards and clinician identity, rather than replacing or undermining them?
4. What does a "unified voice" look like in your EMS system? How could agencies, educators, and leaders align more closely on AI strategy and values?
5. Which of the seven leadership traits listed at the end of the chapter (e.g., curiosity, adaptability, inclusivity) most resonate with your current leadership style, and which require more intentional development?
6. What is one concrete step you or your agency can take in the next 90 days to begin shaping—not reacting to—the future of EMS in the AI era?

Study Questions

1. What does the chapter identify as EMS's most urgent task in integrating AI?

    a. Investing in new equipment
    b. Waiting for federal guidance
    c. Leading with professional values and shared standards
    d. Hiring more software engineers

2. According to the chapter, what historical event marked EMS's recognition as an allied health profession?
3. True / False: The chapter argues that AI alone will determine the future of EMS.
4. The chapter emphasizes that AI should not lead EMS—it must be led by ____________.
5. List three emerging technologies—besides AI—that are mentioned as shaping the future of EMS.
6. True / False: According to the chapter, EMS has historically outpaced other health professions in adopting emerging technology.
7. The chapter outlines several principles to carry forward. Choose any two and explain how they apply to EMS leadership in the AI era.
8. Why does the chapter warn against letting outside entities such as vendors or market players define the future of EMS?
9. How does the story of the 3:00 a.m. night call highlight both the current limitations of AI and the value of human judgment?
10. What is the primary reason the chapter urges EMS to act now on AI integration?

CHAPTER SUMMARY

Embracing the Future of Emergency Medical Services with Artificial Intelligence

Chapter 14 is a strategic and moral call to action for EMS leaders, educators, and clinicians. It frames the integration of AI into EMS not as a distant possibility, but as an urgent imperative. The chapter grounds this argument in historical precedent and contemporary healthcare trends, noting that EMS has often lagged behind other health professions, such as medicine and nursing, in adopting transformative technologies. That gap can no longer be justified.

This chapter argues that AI must be viewed as a permanent part of EMS's future. It can revolutionize clinical care, operational efficiency, workforce sustainability, and patient outcomes. However, realizing this potential depends on leadership—leaders who are curious, collaborative, adaptable, and committed to professional standards. Drawing from examples in other medical disciplines and historical EMS milestones, the chapter urges EMS professionals to unify, modernize training, and build a coordinated national strategy for AI adoption.

Notably, the chapter warns against letting the future of EMS be shaped by outside forces, whether vendors, regulators, or disjointed market players. Instead, it calls on EMS to claim agency over its future through shared standards, bold leadership, and deliberate innovation. Just as the pioneers of EMS once transformed prehospital care into a recognized allied health profession, today's EMS leaders are tasked with ensuring that AI enhances—not erodes—the human mission of emergency care.

Notes

1. The Future Is Now

1. Highway Safety Act of 1966, Pub. L. No. 89–564, 80 Stat. 731 (1966).
2. Emergency Medical Services Systems Act of 1973, Pub. L. No. 93–154, 87 Stat. 593 (1973).
3. "News Items," Academy of International Mobile Healthcare Integration, accessed June 24, 2024, https://aimhi.mobi/news.
4. Idaho Department of Health and Welfare, "Emergency Medical Services Sustainability Task Force," https://healthandwelfare.idaho.gov/about-dhw/boards-councils-committees/emergency-medical-services-sustainability-task-force-emsstf.
5. South Dakota Department of Health, "EMS Sustainability Assessment," https://doh.sd.gov/healthcare-professionals/ems-trauma-program/ems-sustainability-assessment/.
6. Sujata Srinivasan, "Officials look into sustainability of emergency medical operations across Connecticut," *New England Public Media*, July 22, 2022, https://www.nepm.org/2022-07-22/officials-look-into-sustainability-of-emergency-medical-operations-across-connecticut.
7. Ruth Brown, "Bill advances to declare Idaho EMS an essential service," *Idaho Reports*, March 14, 2024, https://blog.idahoreports.idahoptv.org/2024/03/14/bill-advances-to-declare-idaho-ems-an-essential-service/.
8. South Dakota Department of Health, "EMS Sustainability Assessment."
9. League of Minnesota Cities, "League convenes task force to address challenges to emergency medical services delivery and sustainability," https://www.lmc.org/news-publications/news/all/league-convenes-task-force-to-address-challenges-to-emergency-medical-services-delivery-and-sustainability/.
10. American Nurses Association Center for Ethics and Human Rights, "The Ethical use of Artificial Intelligence in Nursing Practice: Position Statement," 2022, https://www.nursingworld.org/~48f653/globalassets/practiceandpolicy/nursing-excellence/ana-position-statements/the-ethical-

use-of-artificial-intelligence-in-nursing-practice_bod-approved-12_20_22.pdf.
11. Nevarda Smith, "Unlocking Potential, Addressing Risks of AI in Medical Practice," MagMutual, June 2024, https://www.magmutual.com/healthcare-insights/article/unlocking-potential-addressing-risks-of-ai-in-medical-practice.

2. Foundations

1. SkyQuest Technology Consulting, "Artificial intelligence (AI) in Healthcare Market to Expand at a Robust CAGR of 38.1% through 2031," *Yahoo Finance*, May 13, 2024, https://finance.yahoo.com/news/artificial-intelligence-ai-healthcare-market-120000102.html.
2. John McCarthy, Marvin L. Minsky, Nathaniel Rochester, and Claude E. Shannon, "A Proposal for The Dartmouth Summer Research Project on Artificial Intelligence," *AI Magazine* 27, no. 4 (2006): 12–14, https://doi.org/10.1609/aimag.v27i4.1904.
3. Casimir A. Kulikowski, "Beginnings of Artificial Intelligence in Medicine (AIM): Computational Artifice Assisting Scientific Inquiry and Clinical Art—with Reflections on Present AIM Challenges," *Yearbook of Medical Informatics* 28, no. 1 (2019): 249–256, https://doi.org/10.1055/s-0039-1677895.
4. Mary Moore and Kimberly A. Loper, "An Introduction to Clinical Decision Support Systems," *Journal of Electronic Resources in Medical Libraries* 8.4, (2011): 348–366.
5. P. Kalaiarasi and P. Esther Rani, "A Comparative Analysis of AlexNet and GoogLeNet with a Simple DCNN for Face Recognition," *Advances in Smart System Technologies*, vol 1163, eds. P. Suresh, U. Saravanakumar, and M. Hussein Al Salameh (Springer Singapore, 2021), 655–668.
6. Alana Holland, "West MI EMS Company Uses AI to Predict Medical Emergencies," WZZM13, April 16, 2024, https://www.wzzm13.com/article/news/health/west-mi-ems-company-uses-ai-predict-medical-emergencies/69-940f05ff-de9a-4bec-a8c0-3fc84536dbe2.
7. Aisling Bracken, Clodagh Reilly, Aoife Feeley, Eoin Sheehan, Khalid Merghani, and Iain Feeley, "Artificial Intelligence (AI)–Powered Documentation Systems in Healthcare: A Systematic Review," *Journal of Medical Systems* 49, no. 1 (2025): 28, https://pubmed.ncbi.nlm.nih.gov/39966286/.

8. Steve Lohr, "Facial Recognition Is Accurate, if You're a White Guy," *The New York Times*, February 9, 2018, https://www.nytimes.com/2018/02/09/technology/facial-recognition-race-artificial-intelligence.html.
9. Dartmouth Summer Research Project on Artificial Intelligence, June 18–August 17, 1956.

5. Health Care's Multibillion-Dollar Bet on Artificial Intelligence

1. Jason Stott, "AI in Healthcare Statistics and Market Trends in 2025," *Open & Affordable*, https://www.openandaffordable.com/post/ai-in-healthcare-statistics-and-trends.
2. Sky Quest Technology Consulting, "Artificial Intelligence (AI) in Healthcare Market to Expand at a Robust CAGR of 38.1% through 2031," *Yahoo Finance*, May 13, 2024, https://finance.yahoo.com/news/artificial-intelligence-ai-healthcare-market-120000102.html.
3. MarketsandMarkets, "Artificial intelligence (AI) Market Worth $1,339.1 Billion by 2030: Exclusive Report by MarketsandMarkets," *PR Newswire*, June 7, 2024, https://www.prnewswire.com/news-releases/artificial-intelligence-ai-market-worth-1-339-1-billion-by-2030--exclusive-report-by-marketsandmarkets-302166982.html.
4. William Van Melle. "MYCIN: A Knowledge-Based Consultation Program for Infectious Disease Diagnosis," *International Journal of Man-Machine Studies* 10, no. 3 (1978): 313–322, https://doi.org/10.1016/S0020-7373(78)80049-2.
5. "Association for the Advancement of Artificial Intelligence," Association for the Advancement of Artificial Intelligence, https://aaai.org/.
6. Ahmet İlker Tekkeşin, "Artificial Intelligence in Healthcare: Past, Present and Future," *The Anatolian Journal of Cardiology* 22 (2019):8–9, https://doi.org/10.14744/AnatolJCardiol.2019.28661.
7. Gordon E. Moore, "Cramming More Components Onto Integrated Circuits," *Proceedings of the Institute of Electrical and Electronics Engineers* 86, no. 1 (1998): 82–85, https://doi.org/10.1109/JPROC.1998.658762.
8. Kat Davis, "Moore's Law Accelerated: The Exponential Growth of AI," *Sidecar*, March 4, 2024, https://sidecar.ai/blog/moores-law-accelerated-the-exponential-growth-of-ai.
9. American Hospital Association, "Digital Health Funding May Be Down, but AI Still Draws Interest from Investors," August 22, 2023, https://

www.aha.org/aha-center-health-innovation-market-scan/2023-08-22-digital-health-funding-may-be-down-ai-still-draws-interest-investors.

10. Sachin Kalotra, "AI in Healthcare: Market Size and Forecast," *Signity Solutions* (blog), September 5, 2025, https://www.signitysolutions.com/blog/ai-in-healthcare-market-size-and-forecast.
11. Emily Olsen, "Healthcare Venture Capital Investment Boosted by AI in 2024: Report," *Biopharma Dive*, January 9, 2025, https://www.biopharmadive.com/news/healthcare-venture-captial-funding-ai-boost-2024-silicon-valley-bank/736902/.
12. Silicon Valley Bank, "Healthcare Industry Trends: 2025 Mid-Year Report," https://www.svb.com/trends-insights/reports/healthcare-investments-and-exits/.
13. Silicon Valley Bank, "Healthcare Industry Trends."
14. American Hospital Association, "Digital Health Funding May Be Down, but AI Still Draws Interest from Investors," August 22, 2023, https://www.aha.org/aha-center-health-innovation-market-scan/2023-08-22-digital-health-funding-may-be-down-ai-still-draws-interest-investors.
15. American Hospital Association, "Digital Health Funding May Be Down."
16. World Economic Forum, "These 5 Countries Are Leading the Global AI Race: Here's How They're Doing It," May 1, 2024, https://www.weforum.org/stories/2024/05/these-5-countries-are-leading-the-global-ai-race-heres-how-theyre-doing-it/.
17. U.S. Congress, House, Committee on Energy and Commerce, "What We Learned: The Change Healthcare Cyber Attack," April 16, 2024, https://energycommerce.house.gov/posts/what-we-learned-change-healthcare-cyber-attack
18. International Business Machines Corporation, "Cost of a Data Breach Report 2023," accessed April 19, 2025, https://www.ibm.com/reports/data-breach.
19. Welsh, Carson, Anderson & Stowe, "ImageTrend Announces Strategic Growth Investment from Welsh, Carson, Anderson & Stowe," *WCAS*, February 9, 2023, https://wcas.com/news/imagetrend-announces-strategic-growth-investment-from-welsh-carson-anderson-stowe/.
20. Chirag Chopra, Ankit Kasare, and Piyush Gupta, "How Venture Capital Is Investing in AI in the Top Five Global Economies—And Shaping the Economy," *World Economic Forum*, May 24, 2024, https://www.weforum.org/stories/2024/05/these-5-countries-are-leading-the-global-ai-race-heres-how-theyre-doing-it/.

6. The President of the United States, Policy, and the National Push for Artificial Intelligence in Health Care

1. The White House, "Public Comment Invited on Artificial Intelligence Action Plan," February, 2025, https://www.whitehouse.gov/briefings-statements/2025/02/public-comment-invited-on-artificial-intelligence-action-plan/.
2. American Medical Association, "Augmented Intelligence in Health Care," 2019, https://www.ama-assn.org/system/files/2019-01/augmented-intelligence-policy-report.pdf.
3. Crigger, E., Reinbold, K., Hanson, C. et al. Trustworthy Augmented Intelligence in Health Care. *J Med Syst* 46, 12 (2022). https://doi.org/10.1007/s10916-021-01790-z
4. American Nurses Association, "The Ethical Use of Artificial Intelligence in Nursing Practice: Position Statement," 2022, https://www.nursingworld.org/~48f653/globalassets/practiceandpolicy/nursing-excellence/ana-position-statements/the-ethical-use-of-artificial-intelligence-in-nursing-practice_bod-approved-12_20_22.pdf.
5. American Academy of Family Physicians Innovation Labs, "Using an AI Assistant to Reduce Documentation Burden in Family Medicine: Evaluating the Suki Assistant," November, 2021, https://www.tafp.org/media/healthy-practice/report-suki-assistant-documentation-burden.pdf.
6. American Medical Association, "Advancing Health Care AI Through Ethics, Evidence, and Equity," https://www.ama-assn.org/practice-management/digital/advancing-health-care-ai-through-ethics-evidence-and-equity.
7. Eric Boyd, "Microsoft and Epic Expand AI Collaboration to Accelerate Generative AI's Impact in Healthcare, Addressing the Industry's Most Pressing Needs," Microsoft (blog), August 22, 2023, https://blogs.microsoft.com/blog/2023/08/22/microsoft-and-epic-expand-ai-collaboration-to-accelerate-generative-ais-impact-in-healthcare-addressing-the-industrys-most-pressing-needs/.
8. Pranavsingh Dhunnoo, "Microsoft's Multi-Tenbillion Dollar Bet On Healthcare," *The Medical Futurist*, November 9, 2023, https://medicalfuturist.com/microsoft-in-healthcare/#.
9. Heather Landi, "HIMSS25: Epic Building Out Agentic AI as The Health IT Giant Also Broadens Focus Beyond EHRs," Fierce Healthcare, March

7, 2025, https://www.fiercehealthcare.com/ai-and-machine-learning/epic-building-out-agentic-ai-it-also-broadens-focus-beyond-ehrs.
10. Landi, "HIMSS25: Epic Building Out Agentic AI."
11. American Nurses Association, "The Ethical Use of Artificial Intelligence in Nursing Practice."
12. Nadia Daneshvar, Deepti Pandita, Shari Erickson, Lois Sulmasy, and Matthew DeCamp. "Artificial Intelligence in The Provision of Health Care: An American College of Physicians Policy Position Paper," *Annals of Internal Medicine* 177, no. 7 (2024): https://doi.org/10.7326/M24-0146.

7. Artificial Intelligence–Powered Documentation in Emergency Medical Services

1. National Emergency Medical Services Information System, "2023 National EMS Data Report," 2023, retrieved April 19, 2025, https://nemsis.org/wp-content/uploads/2025/02/NEMSIS-Annual-Public-Data-Report-2023_.pdf.
2. National Emergency Medical Services Information System, "Monthly Activity Report December 2024," retrieved April 19, 2025, https://nemsis.org/wp-content/uploads/2025/01/NEMSIS-Activity-Report-December-2024.pdf.
3. John McCormack, "As Interest in Ambient AI Grows, HI Professionals Focus on Clinical Documentation," *Journal of AHIMA*, January 6, 2025, https://journal.ahima.org/page/as-interest-in-ambient-ai-grows-hi-professionals-focus-on-clinical-documentation.
4. American Academy of Family Physicians, "Phase Two Trial of AI Assistant for Documentation a Success," December 17, 2021, https://www.aafp.org/news/practice-professional-issues/20211217sukiphase2rpt.html.
5. John Feldman, Katherine Hochman, Benedict Guzman, Adam Goodman, Joseph Weisstuch, and Paul Testa, "Scaling Note Quality Assessment Across an Academic Medical Center With AI and GPT-4," *NEJM Catalyst Innovations in Care Delivery* 5, no. 5 (2024), https://doi.org/10.1056/CAT.23.0283.
6. National EMS Information Systems, "Drug Overdose Surveillance Dashboard," August 13, 2024, https://nemsis.org/nonfatal-drug-overdose-surveillance-dashboard/.

7. Akerman LLP (2026), Texas TRAIGA and California AB 489.
8. Weber (2026, January 16), Saucedo v. Sharp HealthCare.
9. National Emergency Medical Services Information System, "EMS Body-Worn Camera Quickstart Guide: Legal Considerations for EMS Agencies," 2021, https://nemsis.org/wp-content/uploads/2021/06/EMS-Body-worn-Camera-Quickstart-Guide_Legal-Considerations_06.2021.pdf.
10. Community Policing Dispatch, "Using AI to Write Police Reports," U.S. Department of Justice, January, 2025, https://cops.usdoj.gov/html/dispatch/01-2025/ai_reports.html.
11. Fisher Phillips (2025, December 9), Two-party consent litigation theories.

8. Real-Time Decision Support

1. International Academies of Emergency Dispatch, "The Hidden Emergency Crisis: 9-1-1 Staffing is a Challenge in Many Communities," August 15, 2023, https://www.emergencydispatch.org/in-the-news/press-releases/963b3265-c780-447d-a040-053cd7c51958.
2. Ministry of Science and ICT, "5G-Based AI Emergency Medical System Secures the Golden Hour for Emergency Patients!," news release, December 17, 2020, https://www.msit.go.kr/eng/bbs/view.do?sCode=eng&mId=4&mPid=2&bbsSeqNo=42&nttSeqNo=481.
3. Leila Jordan, "Emergency Dispatchers Are Using AI and Cloud-Based Tools to Help Those in Need Faster." *Business Insider*, March 11, 2024, https://www.businessinsider.com/emergency-response-call-centers-using-ai-cloud-technology-2024-3.
4. Fredrik Byrsell, Andreas Claesson, Mattias Ringh, Leif Svensson, Martin Jonsson, Per Nordberg, Sune Forsberg, Jacob Hollenberg, and Anette Nord, "Machine Learning Can Support Dispatchers to Better and Faster Recognize Out-of-Hospital Cardiac Arrest During Emergency Calls: A Retrospective Study," *Resuscitation* 162, (2021): 35–43, https://doi.org/10.1016/j.resuscitation.2021.02.041.
5. Faezeh Afzali, Yunes Jahani, Fatemeh Bagheri, and Reza Khajouei, "The Impact of The Emergency Medical Services (EMS) Automation System on Patient Care Process and User Workflow," *BMC Medical Informatics and Decision Making* 21, no. 292 (2021), https://doi.org/10.1186/s12911-021-01658-9.

6. Erin McBeth, "Reducing Mortality and Saving Money by Identifying Signs of Sepsis Sooner," Epic Share, retrieved April 15, 2025, https://www.epicshare.org/perspectives/streamlining-sepsis-prevention-and-care.
7. John Cull, Robert Brevetta, Jeff Gerac, Shanu Kothari, and Dawn Blackhurst, "Epic Sepsis Model Inpatient Predictive Analytic Tool: A Validation Study," *Critical Care Exploration* 5, no. 7 (2023): e0941, https://doi.org/10.1097/CCE.0000000000000941.
8. Christopher Seymour, Thomas Rea, Jeremy Kahn, Allan Walkey, Donald Yealy, and Derek Angus, "Severe Sepsis in Pre-Hospital Emergency Care: Analysis of Incidence, Care, and Outcome," *American Journal of Respiratory and Critical Care Medicine* 186, no. 12 (2012): 1264–1271, https://doi.org/10.1164/rccm.201204-0713OC.
9. Jonathan Studnek, Melanie Artho, Craymon Garner Jr, and Alan Jones, "The Impact of Emergency Medical Services on The ED Care of Severe Sepsis," *The American Journal of Emergency Medicine 30*, no. 1 (2012): 51–56, https://doi.org/10.1016/j.ajem.2010.09.015.
10. Christopher Seymour, Foster Gesten, Hallie Prescott, Marcus Friedrich, Theodore Iwashyna, Gary Phillips, Stanley Lemeshow, Tiffany Osborn, Kathleen Terry, and Mitchell Levy, "Time to Treatment and Mortality During Mandated Emergency Care for Sepsis," *The New England Journal of Medicine* 376, no. 23, (2017): 2235–2244, https://doi.org/10.1056/NEJMoa1703058.
11. Sacramento County Department of Health Services, "Sacramento County EMS Agency APOT Summit #2 2024," October 3, 2024, https://dhs.saccounty.gov/PUB/EMS/Documents/APOT%20Summit%202024/APOT%20Summit%202%20-%20Presentations%20-%20All.pdf.
12. Nora Eisemann, Stefan Bunk, Trasias Mukama, et al. "Nationwide Real-World Implementation of AI for Cancer Detection in Population-Based Mammography Screening," *Nature Medicine* 31, (2025): 917–924, https://doi.org/10.1038/s41591-024-03408-6.

9. Drones, Driverless Tech, and the Future of Emergency Medical Services Mobility

1. National Highway Traffic Safety Administration, "Automated Driving Systems: A Vision For Safety 2.0" (Report No. DOT HS 812 442), U.S.

Department of Transportation, 2017, https://www.nhtsa.gov/sites/nhtsa.gov/files/documents/13069a-ads2.0_090617_v9a_tag.pdf.
2. Waymo, "Waymo Safety Performance: Rider-Only Crash Rates Compared to Human Drivers," *Waymo LLC*, 2024, https://waymo.com/safety/impact.
3. Mike Dickman, "Will Volkswagen Have a Self-Driving Ambulance Ready By 2025?," Sunrise Volkswagen of Fort Pierce, October 22, 2021, https://www.sunrise-vw.com/will-volkswagen-have-a-self-driving-ambulance-ready-by-2025/.
4. "Peoria, Ariz. Roboride," Beep, https://ridebeep.com/locations/peoria.
5. Mark D. Weinstein, "Students Create Autonomous Ambulance Concepts," *Cedarville University News Releases*, October 11, 2023,
6. "CedarCommons," Cedarville Commons Digital Archives and Repository, https://digitalcommons.cedarville.edu/news_releases/1739.
7. Ani Kelkar, Kersten Heineke, Martin Kellner, Timo Möller, and Robert Brennecke, "Will Autonomy Usher in the Future of Truck Freight Transportation?" *McKinsey & Company*, September 25, 2024, https://www.mckinsey.com/industries/automotive-and-assembly/our-insights/will-autonomy-usher-in-the-future-of-truck-freight-transportation.
8. Sean Fleming, "In Rwanda, Drones Are Delivering Blood to Remote Communities," *World Economic Forum*, December 11, 2018, https://www.weforum.org/stories/2018/12/in-rwanda-drones-are-delivering-blood-to-remote-communities/.
9. Eyewitness News, "FDNY Demonstrates Drones Used in Emergency Situations at First Responder Conference," ABC7, May 5, 2023, https://abc7ny.com/fdny-drone-emergency-situation-life-saving-technology/13215715/.
10. *Flying* (2026, March), Hyde County and Pivotal eVTOL proof-of-concept
11. Federal Aviation Administration (2025), Executive Order 14307 and eIPP establishment.
12. Federal Aviation Administration (2026, March 9), eIPP awardee announcement.
13. North Carolina Department of Transportation (2026, February 26 and March 12), eLIFT-NC proposal and selection.
14. *Flying Magazine* (2026, March), Hyde County and Pivotal eVTOL proof-of-concept.
15. *Flying* (2024, November), Metro Aviation and Beta Alia order.
16. *Flying* (2024, September 4) Beta Granted $20M to Explore Electric Aircraft and Chargers for Emergency Response.
17. Lockheed Martin and Sikorsky (2026, March 23), UH-60MX MATRIX delivery.

18. Lockheed Martin, Sikorsky, and Robinson (2026, March 10), R66 TURBINETRUCK unveiling.
19. John Erich, "Ambulance Safety," EMS World, October, 2010, https://www.hmpgloballearningnetwork.com/site/emsworld/article/10319214/ambulance-safety.
20. Lindsay Graham, Dennis Thomas, Kraig Finstad, Rachel Dennert, and Tyler Southcott, "Analysis of Ground Ambulance Crash Data From 2012 to 2018," (Report No. DOT HS 813 480), *National Highway Traffic Safety Administration*, July, 2023, https://www.ems.gov/assets/Analysis-of-Ground-Ambulance-Crash-Data-From-2012-to-2018_FINAL.pdf.
21. Waymo, "Waymo Significantly Outperforms Comparable Human Benchmarks Over 7 Million Miles," December 20, 2023, https://waymo.com/blog/2023/12/waymo-significantly-outperforms-comparable-human-benchmarks-over-7-million/.
22. Waymo, "Waymo Significantly Outperforms Comparable Human Benchmarks."
23. Autonomous Vehicle Industry Association, "Research & Discoveries (R&D): Waymo Reduces Crash Rates Compared to Human Drivers Over 7+ Million AV Miles," January 17, 2024, https://www.theavindustry.org/blog/waymo-reduces-crash-rates-compared-to-human-drivers.
24. John M. Scanlon, Kristofer D. Kusano, Laura A. Fraade-Blanar, Timothy L. McMurry, Yin-Hsiu Chen, and Trent Victor, (2024). "Benchmarks For Retrospective Automated Driving System Crash Rate Analysis Using Police-Reported Crash Data," *Traffic Injury Prevention* 25(suppl), (2024): S51–S65, https://doi.org/10.1080/15389588.2024.2380522.
25. Vision Research Reports, "U.S. Ambulance Services Market Size, Share, Trends, Growth, and Forecast 2024 to 2033," 2024, https://www.visionresearchreports.com/us-ambulance-services-market/41401.
26. Todd Schneider, "Taxi, Uber, and Lyft usage in New York City," Todd W. Schneider, August 6, 2018, https://toddwschneider.com/posts/taxi-uber-lyft-usage-new-york-city/.
27. Lindsay Graham, et al., "Analysis of Ground Ambulance Crash Data."; Kristofer Kusano, John M. Scanlon, Yin-Hsiu Chen, Timothy L. McMurry, Ruoshu Chen, Tilia Gode, and Trent Victor, "Comparison of Waymo Rider-Only Crash Data to Human Benchmarks At 7.1 Million Miles," *Traffic Injury Prevention* 25, (2024): S66–S77, https://doi.org/10.1080/15389588.2024.2380786.

10. Artificial Intelligence–Enhanced Emergency Medical Services Clinicians

1. The NASA Task Load Index (NASA-TLX) is a widely used, research-backed tool developed by NASA to measure perceived workload across six dimensions: mental demand, physical demand, temporal demand, performance, effort, and frustration. It helps evaluate how taxing a task is on an individual, especially in high-stress or high-complexity environments like aviation, medicine, or emergency response. By quantifying cognitive load, NASA-TLX allows researchers and system designers to optimize performance, reduce errors, and support user well-being. It's particularly valuable in EMS for assessing how new tools or procedures impact clinician workload in the field.
2. Elizabeth Harry, Christine Sinsky, Lotte N. Dyrbye, Maryam S. Makowski, Mickey Trockel, Michael Tutty, Lindsey E. Carlasare, Colin P. West, and Tait D. Shanafelt, "Physician Task Load And the Risk of Burnout Among US Physicians in A National Survey," *The Joint Commission Journal on Quality and Patient Safety* 47, no. 2 (2021): 76–85, https://doi.org/10.1016/j.jcjq.2020.09.011.
3. Agency for Healthcare Research and Quality, "Cognitive Load Theory and Its Impact on Diagnostic Accuracy" [Issue brief], https://www.ahrq.gov/sites/default/files/wysiwyg/diagnostic/resources/issue-briefs/dxsafety-cognitive-load-theory.pdf.
4. Tanya Albert Henery, "How Bad 'Cognitive Ergonomics' Can Drain Doctors' Brainpower," American Medical Association, August 16, 2021, https://www.ama-assn.org/practice-management/physician-health/how-bad-cognitive-ergonomics-can-drain-doctors-brainpower.

11. Ethics, Accountability, and the Emergency Medical Services Artificial Intelligence Framework for Action

1. Ziad Obermeyer, Brian Powers, Christine Vogeli, and Sendhil Mullainathan, "Dissecting Racial Bias in An Algorithm Used to Manage the Health of Populations," *Science* 366, no. 6464 (2019): 447–453, https://doi.org/10.1126/science.aax2342.

2. Sinclair Schuller, Samta Kapoor, Kapish Vanvaria, and Dan Diasio, "Responsible AI Means Finding the Balance Between Risk and Reward," EY, May 7, 2024, https://www.ey.com/en_us/insights/ai/balance-risk-and-reward-with-responsible-ai.
3. Mahshid Abir, Rekar K. Taymour, Jason E. Goldstick, Rosalie Malsberger, Jane Forman, Stuart Hammond, and Kathy Wahl, "Data Missingness in the Michigan NEMSIS (MI-EMSIS) Dataset: A Mix-Methods Study," *International Journal of Emergency Medicine* 14, no. 1 (2021), https://doi.org/10.1186/s12245-021-00343-y.
4. Ariel Guerra-Adames, Marta Avalos-Fernandez, Océane Doremus, Cédric Gil-Jardiné, and Emmanuel Lagarde, "Uncovering Judgment Biases in Emergency Triage: A Public Health Approach Based on Large Language Models," *Proceedings of Machine Learning Research 259*, (2024): 420–439, https://proceedings.mlr.press/v259/guerra-adames25a.html.
5. Michael J. Rigby, "Ethical Dimensions of Using Artificial Intelligence in Health Care," *AMA Journal of Ethics*, 21, no. 2 (2019): 121–124, https://journalofethics.ama-assn.org/article/ethical-dimensions-using-artificial-intelligence-health-care/2019-02.
6. Cansu Y. Elgin and Ceyhun Elgin, "Ethical Implications of AI-Driven Clinical Decision Support Systems on Healthcare Resource Allocation: A Qualitative Study of Healthcare Professionals' Perspectives," *BMC Medical Ethics* 25, no. 148 (2024): , https://doi.org/10.1186/s12910-024-01151-8.
7. Cedars Sinai, "Pursuing the Ethics of Artificial Intelligence in Healthcare," October 25, 2023, https://www.cedars-sinai.org/newsroom/pursuing-the-ethics-of-artificial-intelligence-in-healthcare/.
8. National Institute of Standards and Technology, "Artificial Intelligence Risk Management Framework (AL RMF 1.0)," January, 2023, https://nvlpubs.nist.gov/nistpubs/ai/NIST.AI.100-1.pdf.

12. Teaching Tomorrow

1. Christian Angelo I. Ventura and Edward E. Denton, "Artificial Intelligence Chatbots and EMS: Perspectives on the Implications of Generative AI in Prehospital Care," *Open Access Emergency Medicine* 15, (2023): 289–292, https://doi.org/10.2147/OAEM.S420764.
2. Nathan Eddy, "Medical Schools Train Next Generation of Clinicians to Better Understand AI," *HealthTech*, April 4, 2024, https://

healthtechmagazine.net/article/2024/04/medical-schools-train-next-generation-clinicians-better-understand-ai.
3. Nathan Eddy, "Medical Schools Train Next Generation of Clinicians."
4. Association of American Medical Colleges, "Principles for the Ethical Use of Artificial Intelligence in and for Medical Education," 2025, https://www.aamc.org/about-us/mission-areas/medical-education/principles-ai-use.
5. R. Logan Jones and Karen Anstey, "Foundations in AI Ethics for Medical Students: A Structured Session for Navigating Healthcare's Digital Transformation," Association of American Medical Colleges, 2025, https://www.aamc.org/about-us/mission-areas/medical-education/advancing-ai-resource-collection/foundations-ai-ethics-medical-students-structured-session-navigating-health-care-s-digital.

13. The Strategic Integration of Artificial Intelligence in Emergency Medical Services

1. Emergency Medical Services System Sustainability Task Force, "Emergency Medical Services System Sustainability Task Force: Phase I Report," 2023, https://drive.google.com/file/d/1ZWOHCrducXzcRB3RIut5OAVocYb6EFWx/view?pli=1.

Study Questions Answer Key

Chapter 1: The Future Is Now

1. In the early 1980s, Congress began withdrawing federal support for EMS. This culminated in the Omnibus Budget Reconciliation Act of 1981, which consolidated EMS funding into general block grants to states, dissolved the federal EMS lead agency, and ended dedicated national coordination.
2. B. Hyperlocal and community-driven
3. False
4. Reactive
5. Workforce shortages, financial instability, and professional invisibility
6. Workforce instability, financial unsustainability, unreliable response models, and collapsing volunteer infrastructures
7. The chapter argues that AI should be embraced in EMS leadership because it functions as a force multiplier. AI can reduce administrative burden, support clinical decision-making, and allow EMS to lead innovation rather than react to it. For the first time, the profession has access to meaningful operational leverage. When applied intentionally, AI-driven innovation reshapes how care is delivered, improves integration with the broader healthcare system, and enhances the patient experience.
8. C. Becoming irrelevant in a transforming healthcare system
9. Because those cultural roots influence current attitudes toward change and national coordination

Chapter 2: Foundations

1. Artificial intelligence refers to computer systems that mimic human intelligence, such as pattern recognition, learning, and decision-making. Artificial intelligence is critical for EMS if integrated

responsibly because it can streamline documentation, improve care, and transform operations.

2. C. Virtual Reality Gaming
3. False
4. Voice input
5. The ImageNet Challenge in 2012, where the deep learning model AlexNet achieved record-breaking image recognition accuracy, sparked the modern AI revolution.
6. Example: Life EMS in Grand Rapids deployed a predictive dispatch system that used AI to preposition ambulances, improving response times without adding staff.
7. EMS is currently between the early majority and late majority phases—behind hospitals and insurers, but still capable of catching up with focused leadership.
8. AI hallucinations and algorithmic bias
9. If AI is trained on incomplete or skewed data, it can produce biased or unsafe outputs, misclassifying patients or under-triaging based on demographics, geography, or language.
10. If EMS does not lead in adopting AI, others will define our future, just like Uber redefined the taxi industry by moving faster and adapting sooner than traditional systems.

Chapter 3: Lessons from Emergency Medical Services Visionaries

1. The author adopted a vision-driven, principle-based leadership mindset—focusing on mission, standards, and collaboration rather than being limited by existing resources or conditions. He committed to building something lasting, not just improving what existed.
2. There was no national EMS system, no trained workforce, and no cultural expectations for prehospital care.
3. Money
4. B. Regulation
5. A national computerized EMS reporting system—a vision that predated the technology but inspired NEMSIS.
6. He developed EMT training programs and pushed EMS to be recognized as an allied health profession.
7. It established high-level paramedic training for underserved communities and set the foundation for national EMS education models.

8. AI tools may be fragmented, unsafe, and inconsistent without national standards, undermining equity and operational reliability.
9. They took action despite imperfect conditions, emphasized shared standards, and executed with urgency, just as today's leaders must in navigating AI.
10. Visionary leadership, grounded in principle and executed with discipline, is again required as EMS integrates AI into its future.

Chapter 4: Barriers to Innovation

1. Understanding the root causes of resistance—cultural habits, legacy systems, and regulatory inertia—enables EMS leaders to focus on structural reform, not just tool deployment. Without addressing those foundations, even the best technologies will fail to scale.
2. C. Real-time triage algorithms
3. True
4. Fear masquerading as realism
5. Defensiveness, dismissal, and identity-based resistance—many EMS professionals viewed the idea of autonomous ambulances not as a technical debate but as a threat to their role and control.
6. Such tools may misclassify or misprioritize patients in underserved, rural, pediatric, or demographically diverse settings, leading to inequity, diagnostic errors, and clinical risk.
7. The BlackBerry or iPhone analogy
8. Many EMS regulations were designed for static tools and protocols; they are not equipped to oversee dynamic, evolving AI systems, resulting in delays, ambiguity, or policy paralysis.
9. Anchoring bias, sunk cost fallacy, groupthink
10. Build AI literacy, foster psychological safety, frame AI as a support tool, adopt incrementally, align nationally, and implement systems to detect and correct bias.

Chapter 5: Healthcare's Multibillion-Dollar Bet on Artificial Intelligence

1. Because EMS lacks national alignment, shared infrastructure, and strategic visibility in AI policy and investment spaces. Without unified leadership and readiness, EMS risks being designed around rather than included in AI-enabled healthcare systems.
2. C. Demonstrated operational ROI and clinical performance

3. False
4. EMS lacked institutional proximity to academic medical centers, research funding, and interoperable data systems. While AI matured in hospitals and universities, EMS remained decentralized, underfunded, and structurally disconnected from early innovation pipelines.
5. Moore's Law predicted exponential computing growth. Today, AI capabilities are doubling even faster—every six months—resulting in faster, more innovative, cost-effective tools. This rapid acceleration pressures EMS to act or fall irreparably behind.
6. Ambient documentation to reduce clinician workload; Predictive analytics in ICUs to detect deterioration; AI-driven administrative tools for billing, scheduling, and prior authorization
7. The metaphor illustrates legacy organizations dismissing innovation as a fad and losing relevance. EMS is at risk of the same fate—underestimating AI's disruption and clinging to outdated systems while the rest of healthcare accelerates forward.
8. Healthcare and autonomous vehicles. EMS sits at the intersection of both sectors yet remains underrepresented in funding and strategy. This disconnect must be addressed if EMS is to benefit from or contribute to AI-driven transformation.
9. Fragmented systems and lack of national interoperability; Limited partnerships with academia, vendors, and policymakers to define EMS's AI role (bonus third: Legacy software and vendor resistance to integration)
10. Launch credible, measurable AI pilot programs; establish partnerships with hospitals, universities, and vendors; advocate for inclusion in national policy and funding conversations; shift from reactive to proactive innovation strategies with shared documentation and interoperability standards

Chapter 6: The President of the United States, Policy, and the National Push for Artificial Intelligence in Health Care

1. EMS has not presented a unified, national framework or voice to federal agencies.
2. AI is viewed as a national strategic asset, though policy emphasis varies by administration.
3. AI must enhance, not supplant, clinical decision-making and align with medical ethics.

4. Augmented intelligence
5. Does it work? Does it work for my patients? Does it improve outcomes?
6. Transparency and equity (also acceptable: safety and trust)
7. It is when external entities develop AI tools for EMS without EMS involvement, leading to tools misaligned with EMS needs and values.
8. Microsoft and Epic integrate AI into clinical workflows at scale through cloud infrastructure, predictive tools, and ambient documentation, enabled by open architecture.
9. Develop and promote a unified national EMS framework for AI that includes governance, clinician education, and interoperability standards.
10. If EMS does not lead its modernization, it risks being replaced or sidelined by more adaptive, tech-enabled models, just like the taxi industry with Uber.

Chapter 7: Artificial Intelligence–Powered Documentation in Emergency Medical Services

1. C. High investment in AI-powered systems
2. False
3. Clinician
4. Verbalizing findings ensures that assessments and decisions are captured in real time, strengthening documentation, supporting legal defense, and improving training value.
5. C. Crew scheduling
6. False
7. Intelligent
8. B. Voice-to-text field documentation
9. Outside cameras often lack audio, clinical context, and understanding of EMS practice, leading to distorted interpretations or unfair judgments.
10. C. Accepting that perfect conditions may never arrive

Chapter 8: Real-Time Decision Support

1. To enhance clinical decision-making by providing timely, context-specific insights that reduce error, support judgment, and improve patient outcomes.
2. C. Automated ambulance driving
3. False
4. Ambulance patient offload time (APOT) delays and inefficiencies in EMS-hospital coordination.

5. By deploying connected ambulances using HoloLens augmented reality headsets, real-time telemetry, and AI-driven remote physician support to enhance field care.
6. A. Algorithmic bias and data privacy risks
7. Governance (or acceptable variations: oversight, policy framework, clinical supervision)
8. They provide real-time, multilingual speech and document translation, reducing communication barriers and improving equity in patient care.
9. Identifying early signs of physiological deterioration before visible symptoms appear helps EMS clinicians make proactive care decisions.
10. C. Building systems that support clinician oversight and transparency

Chapter 9: Drones, Driverless Tech, and the Future of Emergency Medical Services Mobility

1. Autonomous vehicles reduce the need for two-person crews, enabling single-clinician models and redeploying staff from driving to patient care. This helps address staffing shortages and burnout.
2. B. High-acuity cardiac arrest responses
3. False
4. The taxi industry underestimated disruption and failed to modernize. EMS risks a similar fate if it delays autonomous adoption and loses nonemergency transport market share.
5. Six
6. B. Smart glasses with augmented reality overlays and drone-based telemetry
7. (1) Autonomous companies can offer more reliable, scalable, and cost-efficient transport. (2) Interfacility transports are predictable, repeatable, and not time-critical, making them ideal for automation.
8. C. They can deliver time-sensitive supplies faster and more directly
9. The U.S. Department of Transportation's National Highway Traffic Safety Administration wrote the *Automated Driving Systems: A Vision for Safety 2.0,* which invites EMS and public safety stakeholders to help shape policy.
10. To ensure EMS leads rather than reacts to change, a technology readiness task force should be formed, local pilot programs launched, regulatory reform advocated, and public stakeholders engaged.

Chapter 10: Artificial Intelligence–Enhanced Emergency Medical Services Clinicians

1. Because EMS clinicians are expected to manage complex, high-stakes situations alone, often without real-time support or sufficient data, which creates unsafe conditions and contributes to burnout. The complexity of modern care demands team-based or technology-augmented support.
2. Cognitive load refers to the total mental effort required to perform tasks. In EMS, intrinsic load (task complexity), extraneous load (system inefficiencies), and germane load (processing and learning) are all elevated due to multitasking in dynamic, high-pressure environments.
3. 87%. According to the Agency for Health Care and Research Quality, up to 87% of diagnostic errors are caused by extraneous cognitive load.
4. False. AI tools are designed to augment EMS clinicians' performance, not replace them. They help reduce mental workload and enhance safety by providing real-time decision support.
5. The author compares AI in EMS to autopilot in aviation—it is used not to replace the pilot but to reduce workload, increase situational awareness, and maintain performance during high-stress conditions.
6. A. This is correct because weather conditions are an external operational hazard rather than a cognitive system stressor, whereas poor EHR interfaces, redundant protocols, and alarm fatigue directly increase extraneous cognitive load during clinical decision-making.
7. Cognitive: The AMA recommends applying cognitive ergonomics to reduce error and burnout by designing systems that minimize unnecessary mental strain.
8. Because AI acts as a "copilot" offloads nonessential tasks, identifies trends or risks in real time, and reinforces human judgment, allowing EMS clinicians to stay focused and confident in critical moments.
9. AI can (1) calculate and monitor clinical indexes (e.g., Shock Index) in real time and (2) provide alerts or prompts for dosing, trends, and protocol adherence, reducing the clinician's need to multitask under pressure.
10. Confidence: The chapter emphasizes that the goal of AI in EMS is to provide assurance and cognitive support so clinicians can perform with greater clarity and reduced error risk, not to automate decision-making or speed up processes at the expense of care quality.

Chapter 11: Ethics, Accountability, and the Emergency Medical Services Artificial Intelligence Framework for Action

1. Incomplete or biased data can lead AI systems to make inaccurate, unfair, or unsafe clinical recommendations—amplifying disparities, misrepresenting patient risk, and automating injustice into care decisions.
2. B. Black patients were systematically deprioritized due to flawed cost-based proxies
3. Explainability means the AI system must present its reasoning clearly and interpretably. Clinicians must understand why the system flagged a condition or recommended an action so they can evaluate it, explain it to others, and make informed decisions.
4. Algorithmic fairness ensures that AI treats all populations equitably and does not reinforce bias. Data accuracy ensures that the input data reflects the real clinical picture. Without fairness, AI can be discriminatory, and without accuracy, AI can be clinically unsafe.
5. The clinician
6. EMS clinicians are accountable for patient care and must critically assess and, if necessary, override AI recommendations to remain the final decision-maker.
7. C. Validity, transparency, and human oversight
8. Continuity of care and fidelity to field conditions
9. Because EMS documentation captures context-rich, real-world details that AI must preserve to support legal protection, QA processes, and hospital coordination. AI must not strip away the nuance of field-based medicine.
10. Integrate AI literacy and ethics into clinician training; demand explainability and equity in AI system design; and establish governance and accountability protocols for AI tools.

Chapter 12: Teaching Tomorrow

1. Historical challenges include integrating new clinical tools (like 12-lead electrocardiograms, intraosseous access, King airways) into already crowded curricula with limited instructional hours and fixed hour-based models. Due to structural limitations, educators often had to choose between breadth and depth.

2. The two primary concerns are
   - *Educational overload:* Curriculum demands continue to grow while instructional time remains static.
   - *Fear of deskilling:* Educators worry that reliance on AI may erode foundational clinical reasoning and decision-making skills.
3. B. Automating credentialing exams
4. In EMS, clinicians often work alone or in pairs, in high-stress environments without access to labs, consults, or EHRs. AI serves as a second set of eyes, offering decision support, differential suggestions, and documentation aid under real-time conditions.
5. False
6. Three key skills include
   - Interpreting and contextualizing AI-generated outputs
   - Recognizing algorithmic bias and understanding system limitations
   - Knowing when and how to override AI recommendations based on clinical judgment
7. Medical schools are integrating AI by embedding it into clinical reasoning, teaching structured reflection (e.g., University of California, San Francisco), offering AI-in-medicine doctoral tracks (Harvard), and aligning AI instruction with service-oriented clinical values (Duke). They treat AI as foundational, not optional.
8. AI literacy
9. C. Transparent and explainable algorithms
10. Delaying adaptation risks falling behind in clinical practice, losing professional credibility, and inheriting systems not designed for EMS realities. Waiting allows others to shape the standards, potentially excluding EMS voices from future policy and operational frameworks.

Chapter 13: The Strategic Integration of Artificial Intelligence in Emergency Medical Services

1. Workforce shortages, financial instability, and escalating complexity of EMS calls.
2. C. By optimizing resource allocation, automating documentation, and improving reimbursement
3. AI reduces cognitive load and documentation burden, improving job satisfaction. It also enhances training with simulation-based learning

and personalized feedback, making EMS clinicians feel more competent and supported.
4. False. AI is designed to support and augment clinician judgment, not replace it.
5. B. By forecasting demand, optimizing deployment, and reducing response times
6. Real-time analysis of EMS data to refine protocols and system performance; AI-powered simulation feedback to improve clinician training and preparedness
7. Reducing human bias, standardizing care protocols, and ensuring clinical decisions are based on objective, data-driven analysis rather than subjective interpretation.
8. C. Clear leadership, ethical governance, and investment in AI infrastructure
9. Cognitive offloading
10. Investment in AI infrastructure and training; ethical and regulatory compliance (privacy, equity, explainability); interdisciplinary collaboration and leadership alignment on goals and implementation strategy

Chapter 14: Embracing the Future of Emergency Medical Services with Artificial Intelligence

1. C. Leading with professional values and shared standards
2. The 1975 recognition of EMT-paramedics as an allied health profession by the AMA
3. False
4. EMS clinicians and leaders
5. Possible answers include: smart documentation platforms; wearable monitors; predictive dispatch systems; virtual care models; language translation apps; autonomous transport systems
6. False
7. Sample responses:
    - *Embrace curiosity:* Encourages leaders to explore new tools and technologies to stay ahead
    - *Demand excellence:* Emphasizes that technology must meet the highest standards of accuracy and equity
    - *Adapt to change:* Reinforces the importance of agility in rapidly evolving environments

8. Without proactive leadership, EMS risks being defined by external forces that do not understand or reflect its realities.
9. The night call shows that human intuition, pattern recognition, and compassion remain essential, even when AI cannot yet see the whole picture. Based on subtle cues, the medic's decision saved a life, illustrating why AI must support, not replace, the human role.
10. C. Delay could lead to being shaped by systems that do not reflect EMS values

Index

B